Acknowledgments

Brigitte M. Gefrich, Geerd Hendel's stepdaughter, and her husband, George Gefrich, of Eliot, Maine, for their enthusiasm and support for this biography. Their careful preservation of photographs and other objects owned by Geerd, and their generosity in sharing them with the author, contributed significantly to the quality and scope of this book.

Special Collections staff at the Raymond H. Fogler Library, University of Maine, Orono, Maine.

Maine Maritime Museum former Chief Curator Anne Witty, Chief Curator Christopher Timms, Collections & Library Services Manager Kelly Page, and archives volunteers, Bath, Maine.

Barbara Rumsey and the volunteer staff at the Boothbay Region Historical Society, Boothbay Harbor, Maine.

Gordon Bok, a friend of Geerd Hendel and well-known folklorist, singer-songwriter, song collector, and woodcarver, whose music reflects his experiences in and around the working boat culture in the Gulf of Maine.

Barbara Dyer, whose employment as an office worker at Camden Shipbuilding and Marine Railway/Camden Shipbuilding/Wayfarer Marine spanned 44 years. Her archival collection at the Camden Public Library's Walsh History Center and her book *Grog Ho* contain a wealth of information about Camden shipbuilding.

Robert Rice, East Boothbay, Maine, who has assembled a comprehensive collection of Rice Brothers Corporation photographs and historic documents.

Alice Hodgdon Thompson, Southport, Maine, of the Hodgdon boatbuilding family, who shared recollections and photographs of Hendel's years in the Boothbay area.

Alden Reed II, editor of *The Waterline*, the newsletter of the Boothbay Harbor One Design Association, for sharing copies of past issues of the newsletter and for text editing.

Dave Tew, Boothbay Harbor One Design owner, who provided important guidance concerning Hendel's all-aluminum sloop *Whistler*.

Helen Stevens Norton for the photographs of the half-hull model of ATR #4.

Kevin Johnson, Photo Archivist, Penobscot Marine Museum, Searsport, Maine.

Dan MacNaughton of Belmont, Maine, a former project manager at Rockport Marine, which builds and maintains wooden boats. At one time, he was Associate Editor of *WoodenBoat* magazine, and his articles have appeared there and in *Maine Boats, Homes & Harbors* and *The Quoddy Tides*. He is chief writer and co-editor of *The Encyclopedia of Yacht Designers*.

Barthold and Ingrid Brümmer, Hamburg, Germany, who shared family photographs of Geerd's childhood, some of the boats he designed, and two half-hull models. Barthold is Hendel's nephew.

Maynard Bray, Brooklin, Maine, a prolific writer who led watercraft preservation at the Mystic Seaport from 1969 to 1975. Since then, he has had a hand in reviving many notable classic boats and in finding caring owners for others. He has spent decades building and restoring boats in his own shop in Brooklin and has been the technical editor for *WoodenBoat* magazine for nearly 40 years. Maynard's kind encouragement, support, and text editing were invaluable contributions to the creation of this book.

Audrey Moody, my spouse and research associate, for her steady encouragement and support in the research and writing of this book and for her extraordinary proofreading skills.

Special appreciation goes to the staff of Maine Authors Publishing—art director Wendy Higgins and graphic designer Michelle Hodgdon for their expertise in the many elements of text formatting and design necessary for creating each page of this book. Thanks also to publishing coordinator Dan Karker for his careful management of the many steps required for the design and publication of this work.

Roger Allen Moody
April 2020

Contents

Foreword

Geerd Hendel's career as a naval architect was both prolific and pioneering. In over fifty years at the drafting table, he had a hand in the design of over 270 vessels—including some of the world's largest and fastest yachts and most beloved one-designs. He produced drawings for everything from boom jumpers to naval destroyers and was an important innovator in the use of aluminum in saltwater. In short, Hendel is an important reminder that Maine shipyards, shipbuilders and boatbuilders, and designers were at the forefront.

Roger Moody's *Of Watercaft Form and Function* is a testament to Hendel's career, a welcome study and curriculum vitae for the vessels that bear his fingerprint. Hendel's path is global in scope and represents a familiar story—the role of immigrants as key figures in American innovation. Born in Hamburg, he attended the Higher Technical Institute in Bremen (Germany, not Maine) before immigrating to the United States. He joined Bath Iron Works as a draftsman in 1929. The timing was unfortunate. Foreign workers, such as Hendel, were the first to be laid off in the stock market crash later that year. But he was rehired in 1932 and ultimately became chief draftsman.

It is important to situate Hendel's work in its original context, during the twentieth-century version of the gilded age; his commissions show that the Great Depression provided scarcity for some and opportunity for others. J. P. Morgan's *Corsair* (IV) was the largest yacht ever built in the United States and launched just months after the Wall Street Crash of 1929. Harold Vanderbilt's *Ranger* was built to defend the 1937 America's Cup in a year when unemployment spiked and national manufacturing fell by 37%. Lead designer Starling Burgess specifically requested Hendel for the project—a testament to Hendel's reputation as a draftsman at the time. All told, he would produce over one hundred plans and drawings for the project.

Hendel's impact is multiplied when one takes into account the various lives and afterlives of the vessels he designed. For example, one of his earlier projects was *Hi-Esmaro,* a sharp 270-foot yacht built at Bath Iron Works for Hiram Manville, president of an insulation and roofing company. The first few years of its life were idyllic; it was the venue for entertaining the Swedish royal family and was the subject of an elegant painting by the Italian artist Luca Papaluca. But it would go on to be converted to a coastal minelayer at the New York Navy Yard in 1940, renamed *USS Niagara*, refitted and reclassified as a patrol boat, and then finally sunk by Japanese aircraft at the Solomon Islands in 1943. In other words, the vessels that Hendel helped design touched nearly every corner of the globe and every facet of society.

Hendel did not simply push tried-and-true methods to their limit, but also pioneered new materials and techniques to be used in shipbuilding and boatbuilding, particularly with his collaboration with the Aluminum Company of America (ALCOA). As Moody explores, Hendel's work on *Whistler*—launched at the Rice Brothers yard in East Boothbay—represented the first aluminum pleasure craft ever built in the United States for saltwater use. *Whistler* became ALCOA's ambassador and was proudly featured in full-page advertisements in *Yachting* and *MotorBoating* during the 1940s. These extol the virtues of aluminum construction that would appeal to any boat owner: "Maintenance is greatly simplified—saving weekends for sailing fun."

Maine Maritime Museum is fortunate to preserve a wealth of material related to this pioneer in the Geerd Hendel Photograph Collection, the Geerd Hendel Naval Architecture Papers, and the Geerd Hendel Drawings collection. But we are more fortunate when researchers such as Moody bring these documents, photographs, and plans into the light. We hope that *Of Watercaft Form and Function* will illustrate for readers the significance of these archival collections, the value of Moody's diligent research, and the importance of Hendel's pioneering work, confirming his station among naval architects.

Christopher Timm
Chief Curator
Maine Maritime Museum
November 2019

CHAPTER ONE

Geerd Niels Hendel: Personal History

Geerd Niels [Nilz or Nils are also found in various sources] Hendel (January 14, 1903–March 30, 1998) was a Maine-based naval architect well-respected for his boat designs and a pioneer in the use of aluminum alloy for masts, rigging, hulls, decks, and cabins.

He was born in Hamburg, Germany, on January 14, 1903, the son of Dr. Justus and Clara (Girardet) Hendel and one of eight children. His stepdaughter, Brigitte Brown Gefrich, describes the family as aristocratic, with their fortune coming from publishing the largest daily newspaper in Germany. Geerd was a sickly child and somewhat insecure. Paternal expectations were that all the Hendel children would progress through a doctorate level of education, yet Geerd did not share that goal. He was educated in public schools and then studied naval architecture at the Higher Technical Institute (state college) in Bremen, which included a two-year apprenticeship at the Deutsche Werft Norbiskrug shipyard in Rendsburg on the Eider River and Kiel Canal in Northern Germany. Hendel received a diploma in naval architecture in 1925.

Deutsche Werft was a German shipbuilding company founded in 1925 when Kaiserliche Werft Kiel and other shipyards were merged as a result of the Treaty of Versailles after World War I, which forced the German defense industry to shrink. The company was owned by the government of the Weimar Republic, and its headquarters was in Berlin. Deutsche Werft started building merchant ships, but when the Nazi Party gained power in 1933, the production was changed to naval vessels.[1]

During Hendel's apprenticeship at the Deutsche Werft shipyard, the yard's design office specialized in steel commercial vessels, but not yachts or pleasure craft. Germany's defeat in World War I temporarily halted the country's export shipbuilding industry, but the company switched production to deep-sea fishing steamers and later, again, to cargo steamers.[2]

In 1927–28, Hendel worked briefly for the renowned German shipyard Abeking & Rasmussen on the banks of the Weser River in Lemwerder. Abeking & Rasmussen had an established reputation for technical innovation, pioneering the use of lighter materials for hull construction and developing improved sound-dampening systems.

After these experiences, Hendel told his parents that he wanted to go to the United States for about two years to explore both the east and west coasts, and he obtained a visa to travel to America.[3] He arrived in the U.S. on November 21, 1928, aboard the Hamburg-Amerika Line's SS *Thuringia*, landing in New York City. In the photo, Hendel is second from the left.

Hendel Photograph Collection, MS-310, Maine Maritime Museum.

He quickly found work in the New York office of naval architect Theodore Wells. After Hendel had been employed for seven months, however, Wells decided to reduce his staff because anticipated design work did not arrive; Hendel was one of those released from employment.

He then applied and was accepted in 1929 as a draftsman at the Bath Iron Works (BIW) shipyard in Bath, Maine. In Bath, he had no car so he walked to work and throughout the countryside. He especially enjoyed walks along the Kennebec River, feeling comfortable with the riverine landscape, which echoed the rivers in his native Germany. He compared the open Maine landscape with the view from his apartment at the corner of106th Street and Broadway in New York, where from his window he could only see another wall, and found Maine vastly more to this liking.[4]

Hendel filed a declaration of intention to become a United States citizen with the U.S. District Court of the Southern District of New York on January 11, 1929, but five years of continuous residency in the United States was required before citizenship could be granted.

Then his employment with BIW ended in 1930 because of the stock market crash, which slowed BIW's workflow significantly. He felt that his lack of U.S. citizenship was also a contributing factor in the termination of his employment at BIW.

During the Depression era of the early 1930s, Hendel worked as a draftsman at the Bethlehem Shipbuilding Corporation in Quincy, Massachusetts, where one of his assigned projects was preparing the construction plan proposal for an 87' fireboat for the city of Portland, Maine. The construction work for that vessel was awarded to East Boothbay boatbuilder Henry Rice, who then recruited Hendel to move back to Maine around 1931 to help with the fireboat construction project.

In an interview with the author on June 9, 2018, Robert Rice, the well-organized and thorough historian for the Rice family's boatbuilding years, recalled that Hendel had acquired a diesel-powered car in the early 1930s when, by necessity, he was living a very frugal lifestyle, but it ran "really rough." Rice remembered that when Hendel parked the vehicle on the Rice Brothers' wharf, the whole structure shook, but Hendel wouldn't cut the ignition for fear the car wouldn't restart.

Hendel returned to work as a draftsman at BIW in 1932 and by 1935 had become their chief draftsman.

In the depression years of 1933–34, Hendel's personal record books at the Maine Maritime Museum archives contain entries indicating that he rented out his 20' sloop and taught sailing lessons in Boothbay harbor for supplemental income.

Hendel's poll tax records at the Maine Maritime Museum show that he was a resident of Boothbay Harbor from 1936 to 1941, living at 23 Roads End. In his 1988 audio tape interview, Hendel recalled renting the third floor of a three-story house at that address, which was near the Fred B. Higgins Lobster Co. (now Bristol Lobster Sales) and from which he had a view of the harbor.

Hendel became good friends with the Hodgdon, Rice, Luke, and Goudy & Stevens boatyard owners, showing obvious interest in whatever boat construction projects were under way. Although characterized by some as a "loner," Hendel and the Boothbay area boatbuilders became fast friends. In a July 26, 2018, interview by the author with Alice Hodgdon Thompson (daughter of boatbuilder Norman Hodgdon Jr.), she noted that Hendel had served as best man at her father's wedding ceremony in 1936 when he and Helen Luke (daughter of Boothbay boatbuilder Paul Luke) were married. Alice later traveled to Geerd's family home in Germany and met his brother Hans.

Alice also noted that her father had provided assistance and coaching to enable Hendel to become a U.S. citizen in 1936; see Hendel's Petition for Naturalization in Appendix II. The petition was also signed by East Boothbay boatbuilder Jacob Arthur Stevens. With his residency requirement fulfilled, Hendel's Petition for Naturalization was approved on January 6, 1936.

After Geerd moved to Camden in 1942, Alice continued the family friendship with him. There, Hendel had met Johanna "Hanni" Buchholz Brown, and they married in 1966; Alice visited the couple at their Camden home in the years that followed.

Because Hendel remained a bachelor until 1966, and unless he was visiting family in Germany, he was often invited to attend holiday dinners with the Stevens family. The theme of spending holidays with area boatbuilders and local friends continued in Camden, where Christmas days were often spent with Cary W. Bok, the vice president of Camden Shipbuilding, and his family.

The creative and inventive naval architect W. Starling Burgess established an office at BIW in the fall of 1934 and needed a draftsman for his boat design projects. Burgess had a racing sloop moored in nearby Boothbay Harbor and had met Hendel there. Burgess was a friend of BIW president Peter Newell and was able to convince him to "loan" Hendel to him for various projects with BIW. In a working relationship that lasted four and a half years, Hendel found the work with Burgess opened new windows of design insight, especially when work for the J-Class America's Cup yacht *Ranger* began in 1936 (see Chapter 3). The use of aluminum alloy in her mast, spar, and hull construction was a key part of Burgess's design work, and in future years Hendel became a champion of the use of aluminum in saltwater environments (see Chapter 4).

Hendel's house at 592 Belfast Road.
Hendel Photograph Collection, MS-310, Maine Maritime Museum.

1990 photo of Geerd and Hanni.
Geerd Hendel Collection, courtesy of Hendel's stepdaughter, Brigitte M. Gefrich.

To support the U.S. Navy's World War II efforts, in 1940 Hendel left Boothbay to work briefly at the William Robinson Shipyard in Ipswich, Massachusetts, preparing design details for 110' sub chasers and for landing barges.

Camden Shipbuilding was undergoing a major expansion in the early 1940s to build wooden ships for WWII, and Hendel was hired there in 1941 as a hull designer. Hendel found Camden to his liking, and in 1943, he purchased a house at 592 Belfast Road (U.S. Route 1) near the Lincolnville town line.

Geerd Hendel remained a bachelor until age 63, when in 1966 he married Johanna "Hanni" Brown of Camden, who was then in her forties. Hanni's daughter Brigitte was 17 years old at the time of their marriage. (Brigitte married George Gefrich and now lives in Eliot, Maine.) Hanni was also a native of Germany and had been widowed when her husband, George Wellington Brown, perished in a 1962 car accident. In 1967, Hendel and Hanni sold his house on Belfast Road and lived at 144 Bay View Street in the harbor-front home that was owned by Hanni and her late husband.

Geerd and Hanni were married for 32 years, until Geerd died in 1998 at the age of 95.

Hendel developed a strong ongoing affection for the Maine coast, its people, and its boatyards, which was sustained throughout his long career working in the state.[5]

Geerd Hendel's Obituary

Geerd Hendel's obituary was published in *The Camden Herald*:

CAMDEN—Geerd Nilz Hendel, 95, died March 30, 1998, at Penobscot Bay Medical Center.

Born Jan. 14, 1903, in Hamburg, Germany he was the son of Dr. Justus and Clara (Girardet) Hendel. He was educated in public schools in Germany, and after two years of shipyard apprenticeship, he attended the Higher Technical Institute in Bremen, Germany, where he specialized in naval architecture, graduating in 1925.

Hendel emigrated to the United States in 1928, and within two weeks earned employment in the office of naval architect Theodore Wells in New York City. After Hendel's seven-month employment with Wells, he was accepted in 1929 as a draftsman at the Bath Iron Works shipyard in Bath, Maine, a position he held for several years.

Mr. Hendel moved to Camden in 1942, where he married Hanni Buchholz in 1966.

He had been a member of the Society of Naval Architects and Marine Engineers since the 1930s and was a member of the Camden Rotary Club.

Mr. Hendel spent more than 50 years behind the drawing board, managing his own firm for the last 30. During his illustrious career, he gained acclaim in America and Europe for his luxury yachts, fishing vessels, and tugboats. He took part in developing the working drawings for the 1937 America's Cup winner *Ranger*, which was made at the Bath Iron Works. He also designed the first aluminum boat in the United States.

Mr. Hendel remained active in his craft until his final months, still receiving inquiries and still showing his designs in yachting magazines. He also shared long friendships with many people, including John Gould and architect Buckminster Fuller.

Surviving are his wife Hanni, his daughter and her husband, Brigitte and George Gefrich of Kensington, NH, his son and his wife, Harald and Ruth Liedtke of Germany, his sister Marianne Brümmer of Germany, his two grandchildren Douglas Schulman and Ragna Liedtke of Germany, and several nieces and nephews, all of Germany.

A funeral service will be held 10 a.m. Friday at Our Lady of Good Hope Catholic Church, 7 Union Street, Camden, with Father Eugene Gaffey officiating. Burial will follow at Seaview Cemetery in Rockport.

If desired, the family suggests a memorial gift to the charity of one's choice.

Arrangements by Laite & Pratt Funeral Home, 9 Mountain St., Camden.

Johanna "Hanni" Brown Hendel's Obituary

Hanni's obituary from *The Free Press*, Rockland, Maine, reads as follows:

Longtime resident of Camden, Hanni Hendel died peacefully at the Homestead in Cushing on March 23, 2013. She was under the care of Hospice. Hanni was born in East Prussia, Germany, and came to the U.S. in 1953. She moved to 144 Bayview Street in Camden in 1954. She married George

Wellington Brown, who died in 1962. He had been in the U.S. Intelligence Service (during WWII) and was a noted international photographer.

Hanni met and married Geerd N. Hendel, a prominently known naval architect in 1966. Geerd died in 1998 at the age of 95.

She is survived by an extended family in Germany, a son Harald Liedtke and granddaughter, Ragna Liedtke. Her daughter Brigitte M. Gefrich and grandson Douglas Schulman survive her in the U.S.

Interment will be private and held at the convenience of the family. Arrangements are with the Long Funeral Home & Cremation Service, 9 Mountain Street, Camden.

CHAPTER TWO

Hendel's Professional Life Chronology

In a career lasting over fifty years, "Geerd Hendel designed an amazing spectrum of power yachts of all sizes, working vessels, tugs, one-design sailboats, sloops and ketches, and the first all-aluminum sailboat built in the United States. While his work generally reflects conservative and classic themes, he had a gift for the harmony of lines in hull form, sheer, and deckhouses which enabled him to depart from traditional strictures and create unusual designs of stunning individualistic beauty. Like many great designers, he knew how to design boats that, good as they might appear on paper, looked their best in the water."[6]

In 1927–28, Hendel worked briefly for the German shipyard Abeking & Rasmussen (founded in 1907 by Georg Abeking & Henry Rasmussen). Abeking & Rasmussen had established a reputation for technical innovation, pioneering the use of lighter materials for hull construction. In the post-WWI era, fine German shipbuilders like Abeking & Rasmussen were able to underbid American yards for boat construction work due to the structure of U.S. tax laws at the time.

The Manhattan-based luxury yacht designers (W. Starling) Burgess, Rigg & Morgan created a brief formal partnership with Abeking & Rasmussen in the early 1920s to take advantage of the exemption on import duties on foreign-built yachts that entered the United States under their own power. It wasn't until the United States' ten percent boat excise tax was lifted in 1926 that stateside yards were again able to bid competitively.[7]

Leaving Abeking & Rasmussen in Germany, Hendel emigrated to the United States in November 1928 and began his career by first working for seven months in the office of Theodore Donald Wells in the naval architecture firm of Herreshoff and Wells in New York City. Wells was a naval architect and marine engineer, and a member of the Society of Naval Architects and Marine Engineers and of the London-based Royal Institute of Naval Architects.

Hendel applied to BIW for work as a draftsman and made the trip from New York City to Bath for an interview. He was hired in July 1929 (see letter of hire in Appendix II) as part of a drafting team of twenty. During 1928–29, BIW experienced a remarkable surge of contracts and was building mostly yachts of 205' to 250' in length, with one vessel being completed every month.

In 1929–30 BIW constructed several yachts that were designed by the New York naval architectural firm founded by Henry J. Gielow, which was a direct consequence of Henry J. Gielow, Inc. having become a significant financial investor in BIW. In his 1988 oral interview, Hendel mentions developing design details for the Gielow-designed yachts *Althea, Hi-Esmaro,* and *Halonia* (see Chapter 7).

A notable yacht designed by Wells during Hendel's tenure at BIW was the steel-hulled, turbo-electric powered *Corsair (IV)*. The ship was the fourth yacht by that name owned by J.P. Morgan and his family. *Corsair* was launched on April 10, 1930, and was the largest yacht ever built in the U.S., with a length-over-all (LOA) of 343' and weighing 2,142 gross tons. Morgan used her for 10 years, mostly on the East Coast and in the West Indies. She was given to the British Royal Navy during World War II and then became a luxury cruise ship on the West Coast until being wrecked near Acapulco in 1949.

Corsair (IV) *was launched into the Kennebec River at Bath Iron Works, April 10, 1930.*

Maine Memory Network, collection of the Maine Maritime Museum.

With the growth provided by his work experience at BIW, in 1929 Hendel applied for and was admitted to membership in the Society of Naval Architects and Marine Engineers, a professional organization founded in 1893 to advance the art, science, and practice of naval architecture, shipbuilding, and marine engineering.

Because the stock market crash in 1929 and subsequent nationwide depression virtually dried up orders for both yachts and commercial work, BIW's contract workflow slowed significantly. By 1931, there were no more orders for the large luxury yachts that had been BIW's specialty in the late 1920s; and because Hendel wasn't yet a U.S. citizen, he was one of the first to be let go.[8]

In the early 1930s, while working as a draftsman at Bethlehem Shipbuilding in Quincy, Massachusetts, one of Hendel's assigned projects was preparing the construction and detail plans for an ocean liner.

The city of Portland, Maine, sought bids for an 87' fireboat, *City of Portland (II)*, of 115 tons, which could pump 6,000 gpm from four monitors (water cannons). The construction work for that vessel was awarded to East Boothbay boatbuilder Henry Rice about 1931, and he recruited Hendel to move back to Maine to assist with construction details of this complex vessel. Hendel worked on the Portland fireboat project for three-and-a-half or four months, becoming good friends with Henry Rice.[9]

Although Hendel was initially hired as a Rice Brothers Corporation employee, and because there wasn't much other work under way at the yard in 1932, he also began—as an independent naval architect—to develop designs for 16' and 18' centerboard boats (see Chapter 4). For several years, he kept

In this December 1936 photo, the City of Portland *(II), termed "Engine Seven" by the department, is operating at the Portland Fish Pier fire. The boat remained in service with the city until it was replaced in 1959.*
Portland Veteran Firemen's Association.

before the Boothbay Harbor Yacht Club leaders the concept of adopting a single racing sloop class, since there was no common denominator among the boats owned by members. (The sequence leading to the development of the Boothbay Harbor One Design Class racing sailboat for the Boothbay Harbor Yacht Club is detailed in Chapter 5.)

The date of Hendel's return to BIW isn't clear from existing records, but it was likely around 1932. At about that time, BIW President Peter Newell submitted bids on the construction of 18 fishing trawlers for Frank O'Hara's Boston-based fishing fleet, and BIW was awarded the contract. Building those trawlers kept BIW alive in the early-to-mid-1930s.[10]

Also in 1932, BIW had gained contracts for the U.S. Navy destroyer *Dewey,* a New York harbor tug for the Turecamo Towing Company, a U.S. Lighthouse Service buoy tender, and for several smaller

W. Starling Burgess

William Starling Burgess (December 25, 1878–March 19, 1947) was an American yacht designer, aviation pioneer, and naval architect. In 1904, Burgess opened a boatyard in Marblehead, Massachusetts, where he designed and built small boats.

About 1908, Burgess had also immersed himself in the aviation business. After he had seen Orville Wright soar, he took flying lessons from the Wright Brothers, licensed their parts and engines, and built his own planes. Near the outset of World War I, Burgess developed the batwing-shaped Burgess-Dunne hydro-aeroplane, which was so stable and easy to fly (by his own account) that it could practically land itself. It found no takers, however. The Burgess Company produced aircraft, float planes, and flying boats. Just before WWI, Burgess sold the business, and during the war, as a Lieutenant Commander, he designed aircraft for the U.S. Navy. Burgess had received the prestigious Collier Award for innovation in 1915, but never made it big as a plane manufacturer. [The Collier Trophy is an annual aviation award administered by the U.S. National Aeronautic Association (NAA), presented to those who have made the greatest achievement in aeronautics or astronautics in America, with respect to improving the performance, efficiency, and safety of air or space vehicles, the value of which has been thoroughly demonstrated by actual use during the preceding year.]

In the summer of 1928, Burgess sailed from yacht club to yacht club on Long Island Sound in a 30-foot prototype racing sloop he called the Atlantic Coast One Design. The purpose of Burgess's one-design creation and cruise was to promote a class of fast boats that were identical for racing and yet could be daysailed as well. Eighty orders for the boat were taken that first summer, and the wooden hulls were built in production-line style at the German shipbuilding firm of Abeking & Rasmussen [where Hendel had been employed in 1927–28]. The class is still active on Long Island Sound.[14]

After the war, Burgess returned to naval architecture and designed many well-regarded yachts. In 1930, he designed the America's Cup defender *Enterprise* and developed it using new technologies, including a mast and struts of duralumin developed by the Aluminum Company of America (ALCOA). He then designed the Cup defenders *Rainbow* in 1934 and *Ranger* (with Olin Stephens) in 1937. *Ranger* compiled a fabulous race record in her first and only racing season that included a 4–0 thrashing of the British challenger *Endeavor II*. *Ranger* is still considered the crown jewel of Burgess's long and distinguished career that was not short on glitter.[15] In 1933, Burgess partnered with Buckminster Fuller to design and build the radical and futuristic Dymaxion car.

In 1935, from his office at BIW, he became a consultant to aluminum manufacturer ALCOA, promoting wider use of corrosion-resistant aluminum alloys for ship construction. At BIW, Burgess developed the aluminum alloy *Alumette* and *Tinavette* test hulls.[16]

vessels. In 1933, BIW's prospects for future success were further enhanced by the award of contracts for two Mahan Class destroyers.[11] The destroyers were 341' LOA, with a beam of 35' feet and a draft of 11', and were powered by two geared steam turbines.

By 1936, Hendel had attained U.S. citizenship and become BIW's chief draftsman.

Although Hendel may have lived in Wiscasset, Maine, in the early 1930s, research by Boothbay Region Historical Society's Barbara Rumsey into Boothbay Region directories and town poll tax and property tax records places Hendel's residence from 1934 until about 1940–41 at 23 Roads End on the eastern side of Boothbay Harbor.

Through his work at BIW, Hendel met the famous naval architect W. Starling Burgess who was assigned an office at the shipyard for his work on the J-Class sloop *Ranger*, and the two men competed in sailboat races in Boothbay Harbor.[12]

Drake Sparkman and Olin Stephens II had formed Sparkman & Stephens in 1928 in New York City, and the company became a legendary naval architecture and brokerage firm. The 1936 *Ranger* design team of the young Olin Stephens and the brilliant naval architect W. Starling Burgess, then about age 58, needed a couple of good draftsmen for the *Ranger* project. Hendel and fellow BIW draftsman Henry Mank were "borrowed" from BIW to work for several months with Burgess at the office Stephens had opened in Bath, assisting with the designs for the new racing yacht (see Chapter 3).[13]

In a May 4, 1936, memo, Burgess confirmed a conversation with BIW President Newell and Vice President Archibald Main that his head draftsman, G.N. Hendel, and draftsman H.W. Mank would be placed on his payroll at rates of $175 per month for Hendel and $24.50 per week for Mank. Both rates were the same as they had been receiving from BIW. The pay rates were based on BIW's 35-hour workweek, however, and Burgess asked that they be adjusted for his longer workweeks. Hendel's résumé (see Appendix II) says that a recommendation for his being hired for the Burgess drafting team came from sailor and naval architect G. Baer Connard, the General Manager at BIW.

From 1935 to 1939, Hendel worked primarily in Burgess's Bath and Wiscasset offices designing yachts and commercial vessels.

Hendel continued to work as a design draftsman for Burgess into 1939, at which point Burgess moved back to New York City.

The bid for the construction of the 1938 lightship *Cornfield*, designed by the U.S. Lighthouse Service, was awarded to Rice Brothers Corporation. Hendel, a resident of Boothbay Harbor during the late 1930s, was again recruited to work with the yard as the complicated systems of this vessel were constructed. Boothbay Region Historical Society volunteer Barbara Rumsey held interviews on May 3 and May 6, 2002, with Jim Stevens, the former owner of Goudy & Stevens boatyard, in which Stevens confirmed that Hendel had come to the Rice Brothers yard in the late 1930s to work on the lightship.

The Cornfield, *lightship LV 118, is 115' LOA with a beam of 26'and a draft of 13.4', had a steel hull, diesel power, two deckhouses, a single lantern mast, and a foghorn. She displaced 412 tons. Electric power for the lantern, foghorn, and extensive radio equipment was supplied by two 10 kW diesel generators. Her hull was painted red, as were the hulls of all lightships.*
Robert Rice Collection.

Cornfield was the last lightship designed by the U.S. Lighthouse Service, and her station was at Long Sand Shoal off Cornfield Point, Old Saybrook, Connecticut, from 1938–57. Renamed *Cross Rip*, from 1958–62 she was stationed at Cross Rip, Massachusetts, near Martha's Vineyard. Between 1962 and 1972, renamed *Boston*, she marked the entrance to Boston Harbor. After about 40 years of neglect, she was restored in 2010 and renamed *Overfalls* in honor of the lightship station closest to Lewes, Delaware, where she is now on display in Canalfront Park.[17]

On May 28, 1939, Hendel wrote to his mother in Hamburg, Germany, that he was working temporarily at the Snow Shipyards, Inc., in Rockland, Maine. There he had been assigned to assist with the construction detail designs of two big fishing trawlers, *Pelican* and *St. George* (see Chapter 13). He was hired there on the recommendation of East Boothbay boatbuilder Jim Stevens.

Also in 1939, with material provided by ALCOA, Hendel had Rice Brothers in Boothbay Harbor build entirely of aluminum alloy a 26' sloop of his own design for his own use. *Whistler* was the first aluminum sailing yacht built in America for the saltwater environment, and Hendel owned and sailed her for over 30 years (see Chapter 6).

To support the Navy's World War II efforts, though, in 1940 Hendel took a job at the William A. Robinson Shipyard of Ipswich, Massachusetts, where he did design work for wood-hulled 110' subchasers (SC's) and small steel landing barges. Four hundred and thirty-eight of these wood-hulled sub chasers were built by small yards on both coasts, and 20 were built at the Robinson Shipyard in 1942–43. Each sub chaser could be constructed in 60 to 120 days.[18]

A U.S. Navy sub chaser (USS SC-1297) *off Cape Henry, Virginia, 17 April 1944. Built at the Robinson Shipyard.*
NavSource Photo Archives, Photo # NH 96489.

In 1941, Hendel worked for a short time for Sparkman & Stephens, the storied firm founded by Olin Stephens and Drake Sparkman in 1929.

Hendel's good friend Jim Stevens moved from Boothbay Harbor to the Camden Shipbuilding and Marine Railway Company in 1941, where he worked as lofting master for the WWII Accentor-Class coastal minesweepers AMc-82 (USS *Governor*) and AMc-83 (USS *Guide*).

Hendel applied to Camden Shipbuilding and was hired as hull designer for the yard, beginning work there on November 22, 1942, and earning $99 per week. From his office in a small building at Eaton's Point, he was put in charge of the hull building department. The yard was producing 165' salvage rescue tugs (ATR's) and 103' personnel transports (APc-class) for the Navy.

In addition, 192' self-propelled, wood-hulled Pine Tree coal barges (MCc-class) were being built for the Maritime Service. The yard was certainly a very busy place, employing about 1700 workers at the time, and Hendel found his primary responsibility was procuring the large quantities of materials needed to construct the hulls of these vessels.

During 1943 and 1944, twelve 165' ATR salvage rescue tugs were built for the U.S. Navy at Camden Shipbuilding. They were of wood construction and powered by steam engines.

Pine Tree coal barge (at right) under construction at Camden Shipbuilding.
Hendel Photograph Collection, Maine Maritime Museum.

In 1945, Hendel created his own independent design firm in Camden and continued to design many boats for Camden Shipbuilding. He moved his office from Camden Shipbuilding to a street-level space at 30 Bay View Street, which overlooked the inner harbor. His landlord was Willard Wight, owner of coal and heating-oil supplier P.G. Willey & Co., which also provided dockage services from their harbor-front location.

In an interview with the author on April 12, 2018, Camden/Rockport resident and sailor, singer, and folk song composer Gordon Bok recalled that Hendel's office was a light, bright office, where his three essential elements of design equipment were always at hand: drafting paper, pencil and ink, and a briar smoking pipe.

Barbara Dyer Collection, Walsh History Center, Camden Public Library.

George I. Dodge Jr., who had studied architectural drawing at the University of Maine, worked as a draftsman under Hendel's supervision from about 1944 to 1947. Camden Shipbuilding office staffer and author Barbara Dyer recalled that Hendel was also at times assisted in his design work by marine draftsman Günther Kleeberg of Belfast, Maine.

Hendel maintained a close association with Camden Shipbuilding throughout his career, and with its master builder, Malcolm Brewer. He also developed strong and lasting working relationships with local one-man master builder Elmer Collemer and naval architect Murray Peterson, producing commercial tugs, fishing vessels, motor yachts, and sailing yachts of all descriptions.

In the late 1970s, Hendel relocated to a small retirement office in his residence at 144 Bay View Street. He continued to actively design until just a few months before his death in 1998 in Camden at the age of 95. His genuine affection for the Maine coast geography, its people, and its boatyards never wavered.

Hendel's office files and design plans were placed by Hendel's wife, Hanni, in the archival collections of the Maine Maritime Museum in Bath, Maine. Although his will directed that they be given to the Mystic Seaport Museum, it also gave considerable discretion to Hanni in its execution. When Maine Maritime Museum curator Nathan Lipfert visited Hanni after Hendel's obituary was published to ask her about his records, she immediately gave them to him to place in the Museum's archives.

Geerd Hendel gained fame as a naval architect in both the United States and Europe for his many innovative designs of luxury yachts, fishing vessels, tugboats, launches, and sailboats. He spent well over 60 years behind the drawing board and the last 30 managing his own firm. His accomplishments were documented over the years in magazines such as *Yachting*, *The Rudder*, and *MotorBoating*, and in books on boat design and boatbuilding.

Geerd Hendel, circa 1945.
Barbara Dyer Collection, Walsh History Center, Camden Public Library.

CHAPTER THREE

The Starling Burgess Years: *Alumette, Tinavette,* and *Ranger*

W. Starling Burgess entered Harvard College in 1897 with the Class of 1901. The 22-year-old Burgess withdrew from Harvard in his senior year, though—before graduating—and opened his own naval architecture design firm in Boston. His deceased father, Edward Burgess, had been a prominent yacht designer of the day, and, like his father, Starling had great mechanical and mathematical abilities and a refined sense of line, form, and spatial relationships. From his mother, Caroline "Kitty" Sullivant Burgess, he received a love of literature and poetry, which he regarded as the foundation for all human accomplishment.

The sinking of the USS *Maine* in Havana harbor on February 15, 1898, resulted in a U.S. declaration of war with Spain on April 11, 1898. Burgess, enlisting in the U.S. Navy, was one of a hundred Harvard undergraduates who volunteered for military service. By special vote of the Harvard faculty, he subsequently received credit for the courses he missed during this period.[19]

Burgess was creative, energetic, daring, and a brilliant mathematician. By 1935, he had achieved the status of an internationally renowned naval architect and had moved from New York City to Wiscasset, Maine, working on various projects for BIW in nearby Bath. (Burgess's life is masterfully chronicled by Llewellyn Howland III in his book, *No Ordinary Being: W. Starling Burgess: Inventor, Naval Architect, Aviation Pioneer, and Master of American Design*, published by David R. Godine, Jaffrey, New Hampshire, in association with the New Bedford Whaling Museum and Mystic Seaport Museum, 2015.)

In 1935, Burgess became a consulting naval architect for ALCOA and from his office at BIW, engaged in work on projects that promoted a wider use of corrosion-resisting metal alloys for ship construction. In a June 1935 letter to ALCOA President Roy Hunt, Burgess "indicates he was to design a coastal motorboat using new corrosion-resistant aluminum for framing and hull and deck plating. The craft would measure approximately 120' on the waterline with a normal displacement of 52 tons and would be capable of making 40 knots in open water."[20]

An additional motivation to move to Wiscasset was that his fourth wife, Anna Dale Biddle "Nanni" Burgess had spent time there in her younger days and had many friends in the area. In fact, Nanni had a home constructed for herself and Burgess at Birch Point in Wiscasset.

Starting July 1, 1935, ALCOA paid Burgess $1,283.33 monthly on a contractual basis, which was to provide compensation for himself plus wages for design draftsman Geerd Hendel and draftsman C.G. Chase. Also, it was to include the cost of office rent, office supplies, etc. Throughout 1936, Hendel remained Burgess's principal design draftsman and, for much of the time, his *only* draftsman at a weekly salary of $38.50.[21]

By 1936, Hendel had been promoted to chief draftsman for BIW.

The office space which BIW assigned to Burgess and Hendel was on the second floor over the U.S. Customs House and Post Office, 25 Front Street, Bath, Maine.
U.S. Government photo.

Aluminum Test Hull *Alumette*

In 1935, prior to his work on the *Ranger* project, Hendel became involved with a Burgess-developed project in which "He [Burgess] proposed that BIW build a hull test section of aluminum alloy to serve as a test bed for various preservative and anti-fouling systems." The recommendation was approved, and Burgess designed the section and his draftsman, Geerd Hendel, prepared the detail drawings. Newell [BIW President Peter Newell] in turn agreed to have the section fabricated at the Iron Works. The "vessel fragment" (humorously called *Bath Tub* by Burgess and *The Old Soak* by those who worked on her), measuring 13 feet in length with a 10-foot beam and a depth of 5 feet, was laid out and fabricated at the Iron Works plate shop; it represented the scaled-down cross-section of a hull measuring approximately 100 feet in length. It weighed 3,800 pounds without ballast (8,000 pounds of ballast was added), was fully riveted, decked with teak, and painted red below the waterline and grey above. Officially named *Alumette,* it even had a formal launching with Mary Newell—Pete's youngest daughter—serving as sponsor and was rolled out of the plate shop on 16 December 1935 decorated with a stub mast from which flew the ALCOA pennant, Burgess's personal pennant at the "bow," and signal flags spelling out "ALCOA." As the collection of ALCOA and BIW representatives gathered, Pete pointed to a protruding piece of metal at the "bow" and told Mary, clutching at the traditional ribbon-wrapped bottle of champagne, to "sock it." She did, and the *Alumette* was lowered gently into the Kennebec for a brief "float" before being placed on a flatcar for the journey to Newport News, Virginia, where she was to be permanently moored.

Virtually unheralded and unknown when she was built, even to most of BIW's 750 employees, *Alumette* was still carrying out her duties as a test section well into the 1960s—vindicating

In his narrative about Alumette *in his masterwork,* No Ordinary Being, *Llewellyn Howland describes Hendel as Burgess's "gifted chief draftsman and design associate."*

Hendel Photograph Collection, Box 1, Folder 5, Maine Maritime Museum.

> ALCOA's faith in the corrosion-resistant qualities of their aluminum alloy. Only periodic painting and the replacement of bronze shaft bearings necessitated by the failure of insulation designed to separate dissimilar metals have since been required.[22]

Hendel noted in his 1988 oral interview that *Alumette*'s propeller and stuffing box were bronze.

Experimental High-Speed U.S. Coast Guard Patrol Boat *Tinavette*

LEFT: 1949 inspection of *Alumette* shows only minimum corrosion after 14 years in salt water. RIGHT: Bilge keel test racks were added in 1948. Reinspected in 1962 and found in excellent condition, the *Alumette* has achieved her original purpose: to prove aluminum's capacity to resist marine corrosion. *Alumette* continues today to be used as a test station to evaluate experimental alloys, weld wire-parent metal combinations and mechanical fasteners. This Alcoa test station is the only one of its kind in the world.

ALUMINUM HULL SECTION, *ALUMETTE:*
A 27-YEAR RECORD OF CORROSION RESISTANCE

An experimental hull built by Bath Iron Works for Alcoa in 1935. Dimensions: 13-ft length, 10-ft beam, 5-ft depth. Weight: 3,800 lb, ballasted with 8,000 lb of lead. Riveted construction. Placed in the water January 1936, at Newport News Shipbuilding and Dry Dock Co.

Purpose: To study the effects of sea water on a variety of alloys, painting systems and dissimilar metal connections. Maintenance was *realistic,* particularly with regard to painting, which was usually done once a year. *Alumette* was inspected over the years by representatives of Alcoa, Newport News Shipbuilding & Dry Dock Co., the Navy, Coast Guard and other Government agencies.

1941 Thorough inspection revealed that all topsides and 90 per cent of the underwater hull were entirely free from corrosive attack. Remaining 10 per cent of underwater hull showed only slight to moderate pitting, which in no way lessened structural integrity.

1945 No changes were noted by inspectors; the *Alumette* was still in excellent condition.

1956 After 20 years of service, some repairs were made. Where insulation at the cast propeller housings was intact, no galvanic attack resulted from the bronze propellers. Where the insulation had failed, galvanic attack did take place and the entire assembly was removed. Some rivets and some decking were also replaced at this time. Hull was in such good shape that it was decided not to retire *Alumette* from service.

It is significant that after 27 years, *Alumette* remains in excellent condition with a remarkably low incidence of repair—and does so while receiving only modest, realistic maintenance. Many useful lessons were learned concerning paint systems and the proper methods of preventing galvanic corrosion. Where experimental paints broke down or were improperly applied, some pitting occurred, but there was no structural weakening. No other economically feasible material could have performed as well.

DESIGNING & FABRICATING
9

ALCOA's report on Alumette *after 27 years of saltwater testing at Newport News, Virginia.*

Manuscript Collection, Marine Historical Association, Inc. G.W. Blunt White Research Library, Mystic Seaport.

Ralph Linwood Snow, in his *Bath Iron Works, the First 100 Years,* relates: "During the next twelve months [1936], Burgess developed the design of a new hard-chine hull form (V form) which promised unusual capacity for high-speed operation and turning capabilities in all types of seas and convinced ALCOA to subsidize the construction of a wooden-hulled 33.5-foot motor (torpedo) boat (model)—named *Tinavette*—to test the new form."[23]

Howland described the 33.5' *Tinavette* as being built of aluminum (not wood), a one-third–scale working model of a proposed 100' high-speed patrol boat for the U.S. Coast Guard. Burgess was to run extensive tests and demonstration trials off Wiscasset. Howland further observed that "For an auto-boat [*sic*] and aviation pioneer, navy veteran, and lifelong lover of both steam and internal combustion, the opportunity to work for ALCOA on high-speed aluminum patrol craft must have been intoxicating. The project not only allowed Burgess to test in real time and with generous financial support some of his most advanced theories of hull shape and performance; it presented him with a light, strong, and still little understood marine construction material with which to do so."[24]

Two towing tank models were developed for testing hull forms. A 1/6 scale model (about 200" or 16') hull of *Tinavette* planked with ½" Philippine mahogany was built by Hutch Roberts of the Hodgdon Brothers boatyard in East Boothbay, with scaling calculations performed by Hendel. The second model was 1/8 scale (about 12', or 144"). The models were tested at the National Advisory Committee for Aeronautics towing tank at Langley Field, Virginia.

Tinavette *on a test run in Penobscot Bay.*

Llewellyn Howland, III. *No Ordinary Being*. David R. Goodine publisher in association with the New Bedford Whaling Museum and Mystic Seaport Museum.

During the summer of 1936, *Tinavette* was put through her paces by Burgess, and a number of naval officers were on board at various times. After one such trip, Commander Ed Brady wrote, "Before clearing the breakwater (Rockland, Maine) he (Burgess) put her through some full-speed turns, and she maneuvers very well and is remarkably stable and has a turning circle of about 75 feet, I should judge. We then went outside Owl's Head and drove her into a good ground swell, and she is very dry and free of pounding at all speeds."[25]

In his 1988 interview, Hendel, with obvious enjoyment, mentioned that Burgess was an early riser and liked to take the boat to Monhegan Island (about 20 miles from Boothbay Harbor) in the early morning hours when ocean waters were calm. With her 300-hp gasoline engine, *Tinavette* was capable of a speed of about 38 knots. The last time Hendel saw *Tinavette,* she was in a shed at BIW and probably was still owned by ALCOA.[26]

Hendel commented that, while working on the *Alumette, Tinavette,* and other design projects, Burgess would sometimes work all night and then take an afternoon off to go sailing in Boothbay Harbor waters.[27]

Burgess developed another proposal for the Navy, a 250' all-aluminum, turbine-powered "light destroyer" incorporating his V-hull design and capable of an incredible top speed of 52 knots (59 mph). "The original idea for this wondership [*sic*] originated in 1935 in the brilliant and fertile mind of William Starling Burgess, one of the Renaissance men of the twentieth century."[28]

"But the project began to stall as Burgess grappled with the construction of the America's Cup defender and the unfamiliar requirements of destroyer design....Perhaps ALCOA was beginning to feel the Burgess project was a bottomless pit into which they poured money, because an abrupt change took place in April; suddenly the Bath Iron Works undertook the responsibility for developing the design detail at its own expense with Gibbs and Cox as its design agent. Apparently, Burgess then served as a design consultant and ALCOA's technical representative to the project at ALCOA's expense.

On November 1, 1937, the Burgess concept design for the all-aluminum high-speed destroyer, as executed by Gibbs and Cox, was presented to the U.S. Navy. The prevailing opinion of the Navy's General Board, however, was that the destroyer was too large to serve as a first-time experimental venture. The concept of aluminum destroyers reappeared briefly in 1941, when the Navy sent a letter of intent to BIW to begin working out the details of a formal construction contract. Orders to build the vessel were finally issued by the Secretary of the Navy in November 1941, only to be countermanded by President Roosevelt a month later when Japan entered World War II."[29]

J-Class *Ranger*

As a naval architecture professional, Burgess had designed the J-Class racing sloop *Enterprise,* which was the America's Cup defender in 1930. She carried aluminum spars and fittings made of 18/8 stainless steel (which meant stronger, lighter, and smaller-diameter rigging), and proved her worth in the duel with *Shamrock V.* In the next America's Cup series in 1934, the Burgess-designed *Rainbow* narrowly defeated the Thomas Sopwith British-designed *Endeavor.*

As the great-grandson of the shipping and railroad tycoon Cornelius Vanderbilt, Harold Vanderbilt was born to great wealth and privilege. After graduating from Harvard Law School, he joined the New York Central Railroad, the centerpiece of his family's vast railway empire, where his father was president.

In August 1936, Harold Vanderbilt invited Burgess to assemble a team to design a new J-class sloop to compete in the 1937 America's Cup challenge. Vanderbilt's invitation, however, included three conditions: (1) that Olin J. Stephens II (of Sparkman & Stephens, who, by 1936 had become the most sought-after yacht designer in the world) be the co-designer; (2) that the co-designers share the design fee equally (it's unclear from records whether the final figure was $10,000 or $12,500 each); and (3) that Burgess cede to Stephens his place in the afterguard (the on-board decision-making members of a boat's racing crew) of the new Cup sloop. Stephens came to Wiscasset a few days at a time to work on the project. The construction overseer retained for the *Ranger* project was G. Gilbert "Gil" Wyland, then employed by the New York Shipbuilding Company, who was to be paid a weekly rate of $76.92. Burgess, Sparkman, and Stephens had agreed to Vanderbilt's conditions, and, recognizing Hendel's skillful performance as BIW's chief draftsman, "borrowed" him from BIW for assistance with the *Ranger* design process. A third draftsman was at times either Henry Mank (from BIW) or C.G. Chase. In 1936, Olin Stephens II opened a Sparkman & Stephens office in Bath, and there he began work with Burgess and Drake Sparkman putting together the working drawings for *Ranger*.[30]

Olin Stephens (left) and W. Starling Burgess.
Hendel Photograph Collection, Maine Maritime Museum.

"By February 1936, Vanderbilt's preparations for a possible new Cup defense had advanced to the point where he could commission Burgess to make some hull, rig, and hardware improvements to *Rainbow*... Among the drawings Burgess and, of course, Hendel produced in their Bath drafting room: plan for keel alteration; plan for outhaul winch on main boom; detail and general assembly of D-4 mast fittings; main boom plans; sail plan…as a small step toward getting back into the business of designing a new Cup defender."[31]

Burgess and Stephens produced four designs between them—coded 77-A through 77-D—and models were made for comparative towing tests in the tanks at the Stevens Institute in Hoboken, New Jersey. The tests demonstrated to Vanderbilt the superiority of Design 77-C, with lines drawn by Burgess, assisted by Hendel.[32]

Howland noted that designing a vessel like a J-Class sloop could entail 100 to 150 ink-on-linen drawings.

BIW built the 135' *Ranger* for Harold S. Vanderbilt as the America's Cup defender of 1937. She proved to be the fastest J-class yacht ever built and easily defeated the British-built challenger *Endeavor II* with seven straight wins in seven races. *Ranger* was known as a "Super J" because she achieved the maximum waterline length that the Universal Rule which governed the J-Boat class then allowed. *Ranger* was designed and constructed to demonstrate the dominance and power of American technology and yachting strength, and she certainly did just that.

Ralph Linwood Snow, *Bath Iron Works: The First 100 Years.* Maine Maritime Museum.

"For ALCOA's [corporate] management, which was paying Burgess a salary to work on the use of aluminum in boats, it had always understood that Burgess might need to take time away from ALCOA business to concentrate on the design of a new Cup sloop. Such work might slightly delay ALCOA's development program; on the other hand, it would draw significant public attention to the use of aluminum in cutting-edge racing yachts. All things being equal, ALCOA and Burgess each stood to gain from such publicity, the value of which may already have been beneficially factored into Burgess's salary."[33]

An analytical report at the Maine Maritime Museum's Hendel archives, dated January 18, 1938, from ALCOA's representative J.C. Chesley to company Vice President and Sales Manager George Stanley, gives this information about the use of aluminum in the three most recent

Ranger.

Geerd Hendel Collection, courtesy of his stepdaughter, Brigitte M. Gefrich.

Ralph Linwood Snow, *Bath Iron Works: The First 100 Years.* Maine Maritime Museum.

America's Cup defenders as provided by Starling Burgess: "The successful America's Cup defenders *Enterprise* (1930), *Rainbow* (1934), and *Ranger* (1937) all carried duralumin masts; but *Ranger* was the first to use (ALCOA) duralumin alloy for all her spars, i.e., mast, boom, and two spinnaker poles. The design of the three successive masts differed a great deal..."[34]

Ranger was also equipped below with lightweight aluminum-frame pipe berths for the crew and with an aluminum "laundry chute" for handling head sails to and from the deck.

Ranger's elliptical mast varied in thickness from .47" at the base to .14" at the head.

Upon completion of his role in the design phase of the *Ranger* project, Hendel returned to BIW.

In late 1935, Hendel again worked for Burgess, who had received the U.S. Navy's announcement of a contest for the design of four vessels: a 54' Motor Torpedo Boat, a 70' Motor Torpedo Boat, and two sub chasers, one of 110' and one of 165'. The two larger vessels were to be made of wood, but the Motor Torpedo Boats were to be designed for aluminum construction.

In his own correspondence of November 29, 1935, Burgess termed the 70' Motor Torpedo Boat as the "70 Ft. High Speed Inshore C.G. [Coast Guard] Patrol Boat." The vessel was to be powered

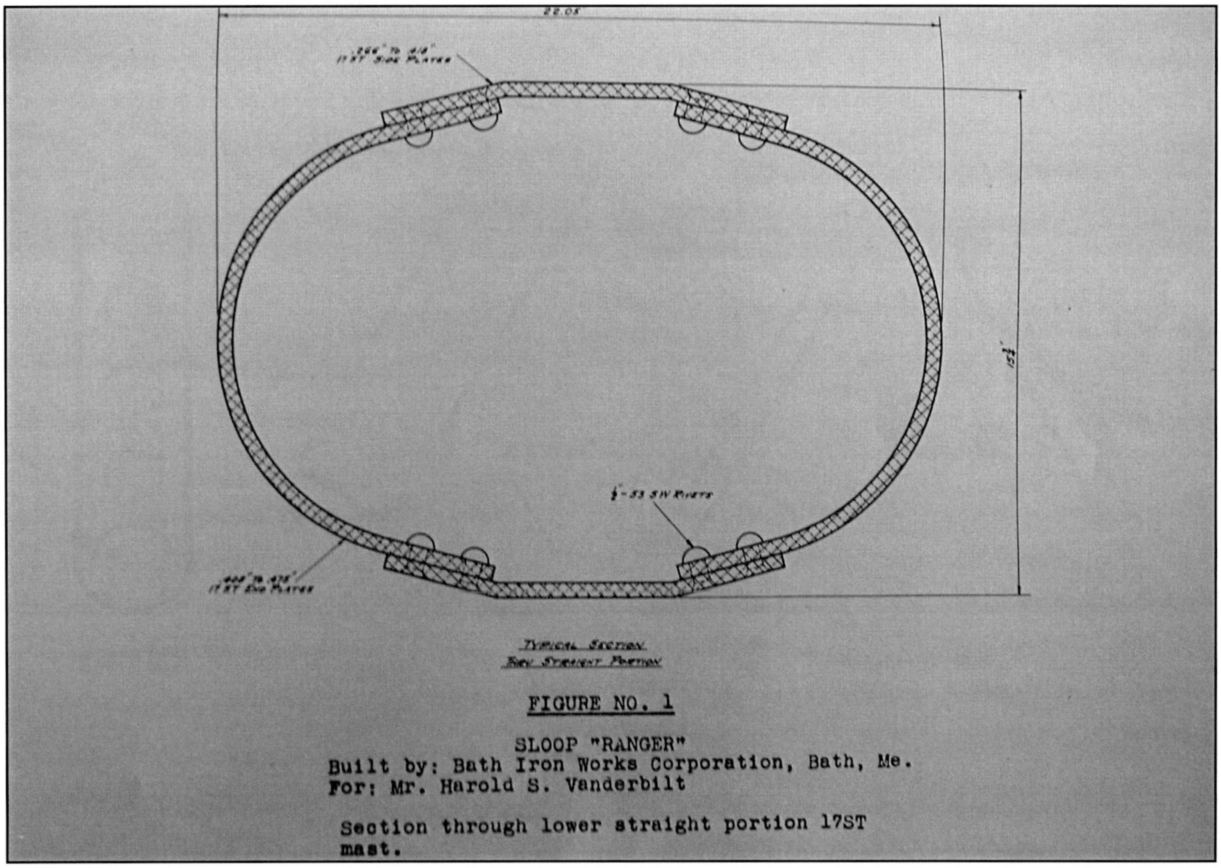

Ralph Linwood Snow, *Bath Iron Works: The First 100 Years.* Maine Maritime Museum.

at 1600 shaft hp maintained by two pairs of 400-hp Liberty motors, each pair being mounted end to end and geared to the propeller shafts. Hodgdon Brothers built 1/6 and 1/8 scale models with Hendel serving as liaison. Burgess directed the model to be towed by the Navy Advisory Committee for Aeronautics test tank facility at a speed corresponding to 55 knots and to evaluate her resistance at lower speeds.[35]

For unknown reasons, Burgess and BIW did not enter a design in this competition. In 1940, the Navy did build an 81' aluminum-hulled torpedo boat (*PT-8*) at the Philadelphia Navy Yard, but it did not incorporate the Burgess-designed V hull and proved to be unwieldy because of the engines chosen to power it.[36]

Burgess's brother Charles, a prominent structural engineer, visited in August 1937, and he taught Burgess, Hendel, and Marjorie Young (who would become Burgess's fifth wife in 1945) a new geometrical method of drawing developable surfaces on either a cylindrical or conical basis. Far from being theoretical, Charles's "new geometrical method" found practical expression in Burgess's design later that year of a pathbreaking 16' V-bottom aluminum centerboarder, *Pirouette.* According to Marjorie (writing as M.G. Young in the November 1939 issue of *Yachting*), "Each side of her bottom is formed of a single sheet of metal cut as if from a huge cone, the curvature beginning with almost flatness at the stern and increasing forward to the full rounded sections at the bow. The topsides are as if cut from a cylinder, their angle with respect to the water being constant." *Pirouette* was built by ALCOA at their New Kensington, Pennsylvania, facility in the spring of 1939 and sailed during that summer season (see Chapter 6 for more detail).[37]

Burgess also designed an aluminum Eight-Meter Class sailing yacht in late 1937, although it is not clear whether the eight-meters were built or remained a design. An aluminum launch to serve the 103' bronze Herreshoff yawl, *Thistle*, was designed by Hendel. On June 8, 1938, the gasoline-powered aluminum-alloy launch, 15' long with a beam of 4.5' and a draft of 2', was launched by BIW [hull #179] as the yawl's tender.[38]

Hendel's working relationship with Burgess ended when Burgess moved his office from Wiscasset back to Manhattan in 1939.

Hendel and Aluminum Construction

In 1939, Hendel designed for his own use the 26' aluminum-hulled *Whistler*, which was built at the Rice Brothers yard in nearby East Boothbay with aluminum supplied by ALCOA. Hendel sailed her for more than 30 years (see Chapter 6).[39]

A Hendel-authored article titled "Aluminum Construction," which was subtitled "Of the Numerous Boat Building Materials, Aluminum Is One of Great Possibilities," was published in the September 1947 issue of *Yachting* magazine.

> In these times of industrial progress, it is small wonder that even the yachtsman and small boat owner is looking for improvements in his craft. Beyond doubt, the various woods, native and foreign, have served him well since the time boats were built for the pleasure of sailing and relaxation, and long before that in commercial shipping. The feel of wooden ships—and the looks of them, too—has imbedded [*sic*] itself into the subconscious mind of man, and he has come to love them and accept them as standard. Most commercial vessels have grown to sizes where wood is no longer feasible. Not so the average yacht. Yet, like most manmade creations, a vessel, particularly a wooden yacht, has, considering the amount of painstaking work we are devoting to building her, a relatively short life span. During the sailing season, sun, wind, and water are at work continually, nor does the drying-out period during the inactive season help matters any. Depending on the quality of material used and, more important, the kind of care her owner is willing and able to give, 20 to 35 years of useful service is about all that can be expected. Though one does not build a yacht for eternity, these facts seem to justify the search for a boatbuilding material which may possibly extend the years of usefulness and, more important, simplify care and upkeep. Checking over likely materials, we find that metals appear to offer great possibil-

ities. Outstanding among them are aluminum as well as the steel alloys. The latter have been ably discussed on these pages and we want to take a look at the newer of the two, aluminum,* particularly with regard to boats larger than dinghies and small factory-built craft.

* Wrought aluminum alloys which are recommended for marine use are designated as "magnesium-silicide." They contain silicon, magnesium, and chromium in varying minute quantities according to the purpose for which they are intended. In addition, the alloy called 618T contains a trace of copper. The suffix T indicates that the material has been treated. The casting alloys 214 and 356 contain magnesium but no copper or chromium. Number 356, which must be heat treated, also contains a substantial amount of silicon. This alloy is intended for uses where strength is important, Alloy 314 is for low-stressed parts.

Among the various aluminum alloys offered to the trade, some are useful to the boatbuilder, others distinctly not. By useful, we have chiefly in mind saltwater corrosion-resisting qualities, all-important deciding factors, as well as workability. At the conclusion of World War II, the U.S. Government offered huge surplus stocks of aircraft aluminum which, for the major part, were not suitable for boat construction. Most high-tensile aluminum alloys as used in planes contain certain quantities of copper, corrode freely in contact with salt water and, therefore, could be used only above water, and must be well protected by paint. Checking the list of available alloys, one may safely select the following: 52S and 61ST for all plate work, 61ST for framing and rails, 53ST-T61 for rivets, and 214 or 356-T6 for castings. They combine strength and workability with excellent resistance to corrosion.

It is surprising how deeply fear of corrosion is rooted in the minds of some persons with whom we come in contact while designing aluminum boats. Usually, the cause is some unfortunate experience in the past, when suitable alloys were not available and just "aluminum" had been used with sad results. Frankly, corrosion is not a problem if the proper alloys are chosen. Saltwater evaporation on an unpainted aluminum deck, particularly in warm, humid weather, can start corrosion. On the other hand, the flow of water around the bottom plates, especially if cold, causes virtually no reaction on the metal. Corrosive action is quick but does not spoil the surface unless allowed to continue for months. The whitish oxide film which develops seems to slow the process but hardly looks attractive; we should protect the metal. A coat of common floor wax rubbed on will do it and preserve the natural color of the aluminum. Yet, because of the poor wearing qualities of wax in wind and weather, it would be most impracticable, and common marine paint of any color desired should be applied. Paint adheres well to aluminum, provided the metal surfaces, with special attention to weldments, have been cleaned thoroughly with chemical dirt remover prior to the application of priming coats. Paint on a metal deck of any kind wears quickly, for which reason all surfaces to be walked on should be sheathed with wood, rubber matting, or other suitable materials.

> The composition of the alloy must be selected with proper regard for the conditions of service in which it will be subjected, but buying "aluminum" will not do.
>
> More serious attention has to be paid to the matter of electrolysis caused by dissimilar metals directly connected to each other and submerged in salt water....To eliminate this condition, we must insulate motor and shafting as well as all other non-aluminum metals completely from the aluminum hull with non-metallic materials.[40]

The article further explored the causes of and treatment for electrolysis of aluminum caused by non-aluminum metals directly connected to each other without being properly insulated. It also reviewed the common boatbuilding tools that could be used for building aluminum boats. The use of aluminum masts and spars was discussed, noting that one-piece extruded spars up to 35' were then commercially available.

Hendel observed that owning an aluminum boat was like owning a wooden boat, with regular maintenance required. Aluminum gives a boat a different personality, but ultimately the experience on the water remains the same.

CHAPTER FOUR

Centerboard Racing Sloops/Boothbay Harbor 20

In the early 1930s, Hendel designed a centerboard sloop called the "Maine Camp One Design" for Camp Wabanaki, a boys' camp in the Hillside area of the town of Sebago, Maine. It was a safe boat, 15' LOA with a beam of 5'3", and carried 100 square feet of sail. Two were built by Rice Brothers in East Boothbay in 1932, and one was built at the Mt. Desert Yacht Yard at an unknown date. The Maine Camp One Design plan below was drawn by Hendel at Camden, where he moved in November 1941, and might be the plan for the boat built at Mt. Desert.

In October 1932, a 16' "one design" class centerboarder designed by Hendel was launched at the Goudy & Stevens yard in East Boothbay, the first of this class to be built.[41]

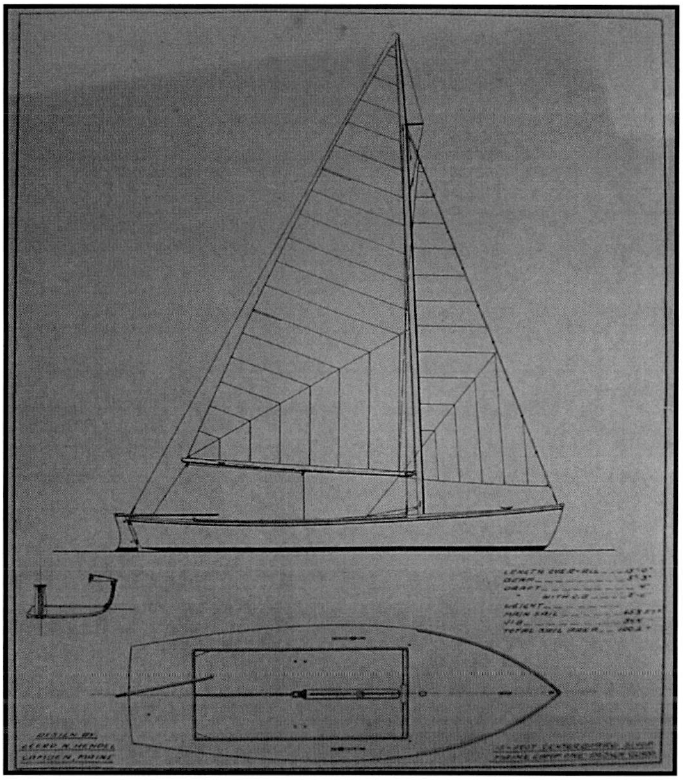

Maine Camp One Design.
Geerd Hendel Collection, Box MS-310, Maine Maritime Museum.

German 12-Meter Centerboard Sharpie's Influence on Hendel's Designs

It's clear that Hendel's forthcoming designs for 18' and 20' centerboard sloops evolved from the early 1930s German lake-racing boat designs, such as the very successful German 12-meter Sharpie, which was 19.5' LOA and had a beam of 4.7'. (In Europe, the Sharpie was a simplified box shape of the round-

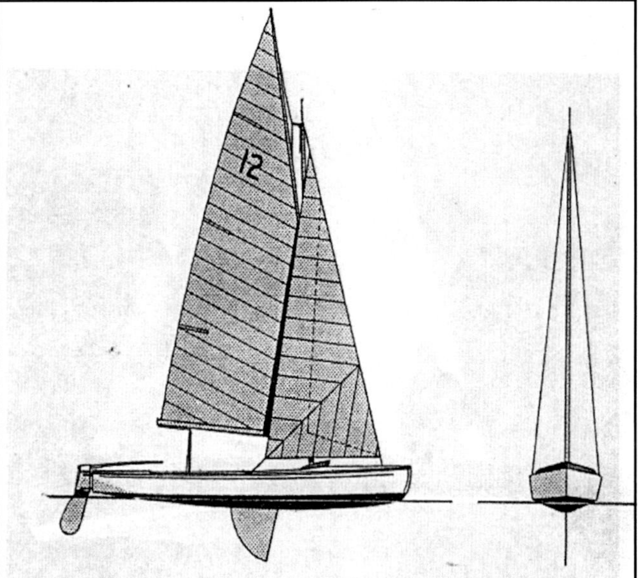

The plan of the Sharpie shows the strong construction, straight lines and fine bow. Illustrations from the 1954 book "British and International Yacht Racing Classes".

hulled Jollenboot class, which was set up with restricted sail area and a box rule.) As a youngster, Hendel had occasionally rented these boats for lake sailing.

The German Sharpie carried a gunter rig with a minimal area of 129.16 square feet (12 square meters) boosted to an actual area of 16 square meters by the big overlap on the genoa. In an amazingly short space of time, the design outgrew Germany and became the world's most widespread performance dinghy.

Shown here are a plan of a German Sharpie with a centerboard and gunter rig, and an image of a German Rennjolle sharpie under sail.

Geerd Hendel Collection MS-310 38.106.1,
Maine Maritime Museum.

This 18-footer, completely rigged and ready to sail, can be purchased for $500, but the price is subject to change without notice. Cradle and shipping expenses extra. As boats will be built to order only, slight changes in the construction to suit individual requirements may be allowed without extra charge. In general the date of delivery will be four to five weeks.

It is worth mentioning the fact that larger boats of this centreboarder type also make very useful crafts on inland waters with shallow shores. For example a boat 30 feet long, 8 to 9 feet wide and 12 to 18 inch draft can be equipped with a surprisingly roomy cabin, including galley, toilet, etc. Aft and just back of a self-bailing cockpit an outboard motor could be installed with the new Harvey inboard motor case. The motor is hinged to the cast bronze or cast aluminum alloy case in such a manner, that it can be tilted either inboard or out below the keel for driving the boat under power. This arrangement makes it possible to sail tne boat with tilted in motor and perfect flush bottom. In other words, you can sail this auxiliary yacht without the resistance of wheel and struts below the keel, that handicaps speed and sailing qualities of all other auxiliaries. It's a most astonishingly simple "inboard" motor arrangement, noiseless and cleanly operating. A common 12 H.P. outboard motor would drive a boat of the size mentioned above at about 8 miles an hour.

These larger and able centreboarders possess the same excellent sailing qualities as the smaller crafts of this type. They are yachts of distinguished appearance and perfect cruising boats for coast and inland waters.

We shall be pleased to submit specifications that fit your requirements satisfactorily in every way.

GEERD N. HENDEL

NAVAL ARCHITECT

BOOTHBAY HARBOR, MAINE

Eighteen Foot Centerboard Sloop

DIMENSIONS

Length, o. a.	18 ft. - 0 in.
Beam	5 ft. - 3 in.
Freeboard (lowest)	16½ in.
Draft	5 in.
Draft with centreboard	2 ft. - 10 in.
Mainsail	93 sq. ft.
Jib	37 sq. ft.
Total sail area	130 sq. ft.

An able boat, that makes the ideal outfit for waterfront cottages and camps. Although sturdily constructed and neatly built in every part, she's light enough to be a fast and sporty craft, just the right boat for yacht club one-design-class races. With fair sheer, a nicely shaped round bottom hull and modern Marconi rig, this 18-footer certainly is the most attractive boat of her kind to be sailed on coast and inland waters.

The specifications read as follows: One-half inch cedar planked hull, copper fastened and plugged. Keel, stem, sternboard, floors, steam bent timbers and deck beams are of selected oak and of fitting proportions. Cockpit floor and deck are of pine and the latter is covered with canvas and sufficiently supported. All trimmings, such as fender, cockpit riser, rails on deck, etc., are of mahogany. The centreboard trunk is of oak, strongly constructed and braced by six cast bronze knees to the floors. The centreboard, a stiff plate of rolled bronze, is served by a manila rope tackle, which is located between mast and centreboard trunk and leads to a cleat at the rear of the trunk. Pulled up, the centreboard does not project below the keel. The rudder blade, also a bronze plate, is hinged into an oakwood rudderhead in such a manner that it tilts back and aft if touching ground. The rudderhead is hinged to the sternboard, and a handy double tiller made of steam bent oak leads over the deck.

Hendel's 1932 18' centerboard sloop selling for $500.

Geerd Hendel Collection, Box MS-310, Maine Maritime Museum.

The next iteration of Hendel's racing sloop designs in 1932 was this 20'4" gunter-rigged centerboard sloop, which came to be known as the Boothbay Harbor 20.

The 20'4" centerboard sloop design clearly reflected the heritage of the 12-square-meter Sharpies popular in Germany for lake sailing, as described in the March 1932 issue of *The Rudder* magazine, p. 76.

No benches are built into the large cockpit, because cushions, laid out on the level floor-covers, make the most comfortable seating. The floorspace in this cockpit is unusually large and extends far under the deck. Aft of the centre-board trunk a flush hatch in the floorcovers is provided for, so that the bilge can be drained with a sponge to the last drop.

The room below deck, forward and aft of the cockpit, can be used for stowing away spare ropes etc. It also contains two galvanized air tanks. These tanks are capable of keeping the boat afloat in case it becomes filled with water and loaded down by several persons.

The mast is of solid spruce and fitted with a sailtrack. For the sails a light yacht duck is used. The mainsail carries four batons, none over two feet six inches in length, and is laced to the boom. The forward end of the boom is fitted with a reefing gear of cast bronze, smooth, simple and unbreakable, making it possible to roll the sail around the boom. With this device and a split, cast bronze ring to carry the mainsheet on the boom, one person staying beside the mast, rolls the boom into the sail as high as desired. Then, the halyard is to be lowered until the boom is in its old position and the mainsail is perfectly reefed. Taking out the reef necessitates nothing more than setting the mainsail higher up again and rolling the boom down and out. It's all so easy and can be done by one person in a minute without assistance.

The lower mainsheet block is fitted to a strong beam running across the cockpit.

The large jib overlaps the mainsail considerably. It is attached to the jibstay by snaphooks and is double sheeted. The sheets run to fair-leaders on the deck and through holes in the cockpit riser to conveniently located cleats.

All manila ropes are of 5-16 inch diameter, while jibstay and shrouds are 1-4 inch diameter galvanized wire rope. The turnbuckles are of bronze, and most of the other fittings, such as two bow chocks, cleats and blocks, are of galvanized, maleable iron. All fittings are of useful size.

Deck and hull are painted outside two coats, and the latter also is sandpapered thoroughly before and after the first coat of paint. Inside, one coat of varnish is applied and the floorboards are painted. Ochre is used for the deck, ivory white for the hull above, and dark-red marine bottom paint below the waterline. All mahogany trimmings, mast and boom have two coats of spar varnish. However, colors may be chosen to suit the individual taste without extra charge.

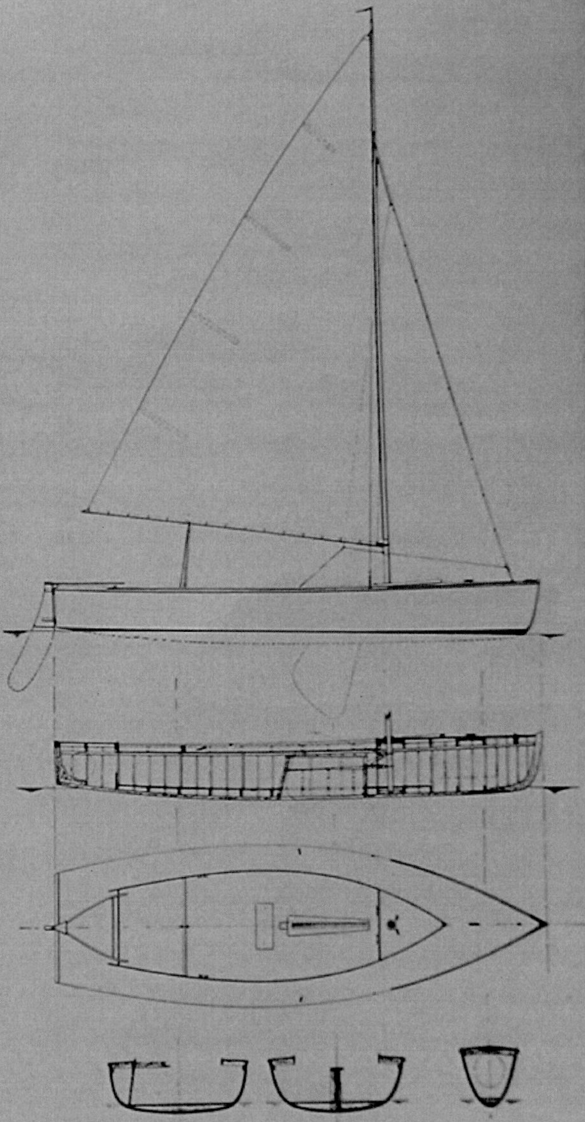

It must be mentioned that this 18-footer is in general but a slightly reduced copy of our 20-foot centreboarder. However, the boat developes the same excellent sailing qualities as well as speed, pointing high to windward. It is seaworthy, is easily sailed single handed, and yet has plenty of room to carry a full crew through snappy races. After having served you for a delightful season on the water, this handy little craft won't cause difficulties concerning winter storage. Pulled up on a beach and carried into a barn is all the storage attention this boat requires in exchange for more than a decade of pleasant, healthy sport.

Hendel's 1932 18' centerboard sloop brochure, inside page, shown with a masthead sail rig.

Geerd Hendel Collection, Box MS-310, Maine Maritime Museum.

Hendel's 20'4 centerboard sloop ad from Yachting *magazine, January 16, 1932.*
Geerd Hendel Collection, Box MS- 310, Maine Maritime Museum.

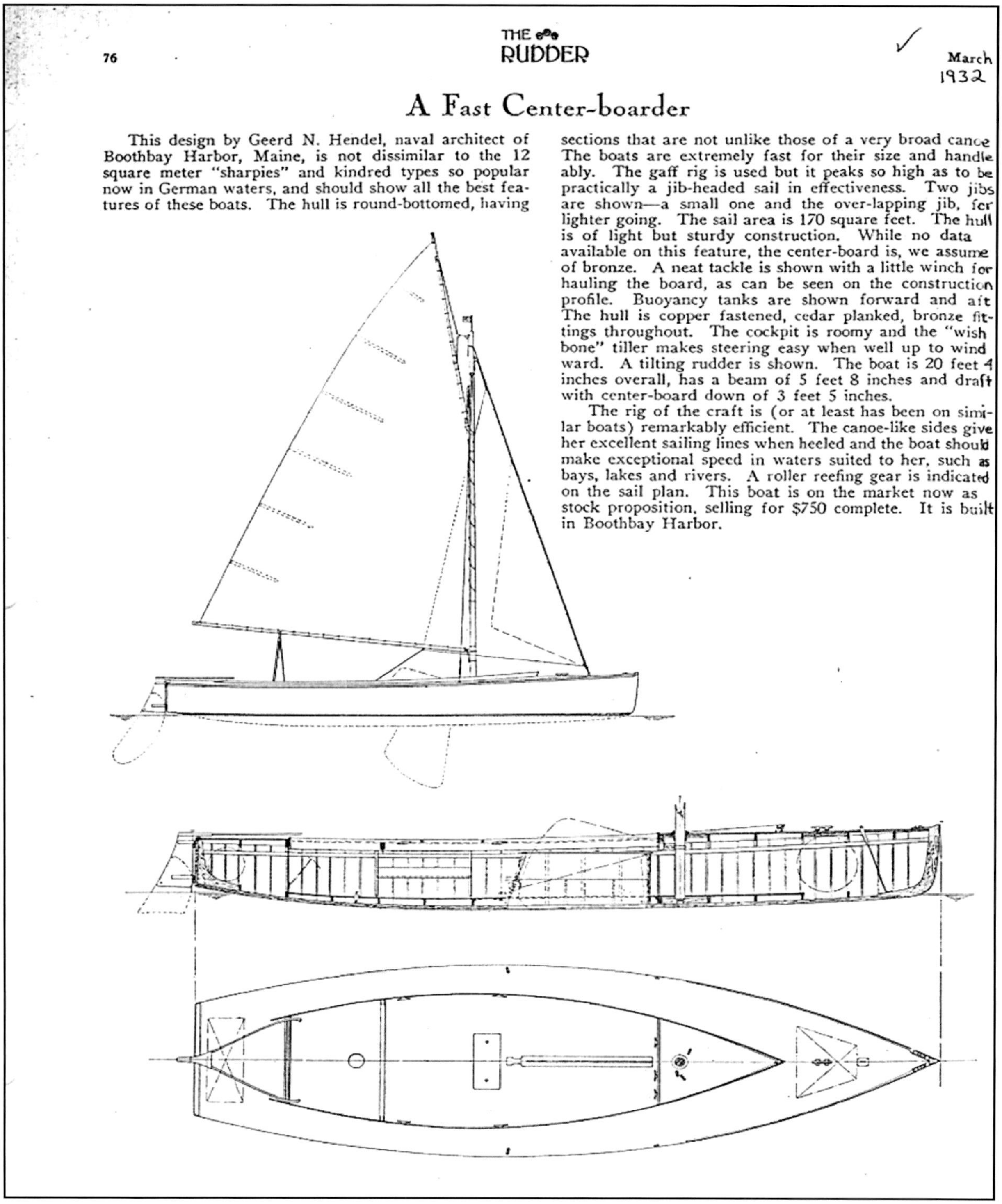

A Fast Center-boarder

This design by Geerd N. Hendel, naval architect of Boothbay Harbor, Maine, is not dissimilar to the 12 square meter "sharpies" and kindred types so popular now in German waters, and should show all the best features of these boats. The hull is round-bottomed, having sections that are not unlike those of a very broad canoe The boats are extremely fast for their size and handle ably. The gaff rig is used but it peaks so high as to be practically a jib-headed sail in effectiveness. Two jibs are shown—a small one and the over-lapping jib, for lighter going. The sail area is 170 square feet. The hull is of light but sturdy construction. While no data available on this feature, the center-board is, we assume of bronze. A neat tackle is shown with a little winch for hauling the board, as can be seen on the construction profile. Buoyancy tanks are shown forward and aft The hull is copper fastened, cedar planked, bronze fittings throughout. The cockpit is roomy and the "wish bone" tiller makes steering easy when well up to wind ward. A tilting rudder is shown. The boat is 20 feet 4 inches overall, has a beam of 5 feet 8 inches and draft with center-board down of 3 feet 5 inches.

The rig of the craft is (or at least has been on similar boats) remarkably efficient. The canoe-like sides give her excellent sailing lines when heeled and the boat should make exceptional speed in waters suited to her, such as bays, lakes and rivers. A roller reefing gear is indicated on the sail plan. This boat is on the market now as stock proposition, selling for $750 complete. It is built in Boothbay Harbor.

Records show that about two dozen of these 20'4 centerboarders were constructed between 1932 and 1942 at the Rice Brothers yard.

Shipyard Collection #5, Rice Brothers, Box 2, Vol. II 1925–42, Boothbay Region Historical Society.

CHAPTER FIVE

Boothbay Harbor One Design and Sister Racing Sloops

With some of the most talented yacht designers and boatbuilders of the early 1930s living in the Boothbay region, like summer visitor L. Francis Herreshoff and area residents W. Starling Burgess, Geerd Hendel, and Charles Hodgdon, it's no surprise that they went to work at their drawing boards to design a sailboat class for local racing.

In 1932, Hendel sparked local interest in the development of a one-design racing sloop for the Boothbay Harbor Yacht Club (BHYC) with the introduction of a class of centerboard boats 20'4" in length with a sail area of 190 square feet (see previous chapter). Hendel, Burgess, and Herreshoff consulted informally and made some adjustments to Hendel's model, and the East Boothbay boatyard of Goudy & Stevens built four of these centerboard hulls in 1934.

Brutal Beast was a class of hard-chine catboat centerboarders designed by W. Starling Burgess in 1912, 14' LOA with a beam of 6' and carrying 120 square feet of sail on a gunter rig. (It's said that the "Brutal Beast" name came from a Great Dane dog owned by Burgess, who was gentle in spirit but ferocious in appearance.) Six of the early boats were for the Portland Yacht basin, and well over 100 were sold to Massachusetts and Maine owners over the years, especially for training young racing skippers.

Boothbay Region Historical Society curator Barbara Rumsey, in notes from a spring 2002 interview with East Boothbay boatbuilder Jim Stevens, conveyed local sailor Phil Haselton's (owner of the 18' centerboarder *Hardalee*) opinion that the first Hendel 20' boat built in the Boothbay region was likely the 1932 *Lao-Tse*, a fast centerboarder built for the Reverend Thomas W. Attridge. Although Haselton felt *Lao-Tse* could probably be regarded as a prototype of the BHOD like the other 1934 20' centerboard boats *Red Spot*, *Medrick*, *Windward* and *Leeward* (all built by Goudy & Stevens), *Lao-Tse* doesn't appear in any other research or commentary about the progression of vessels leading to the BHOD.

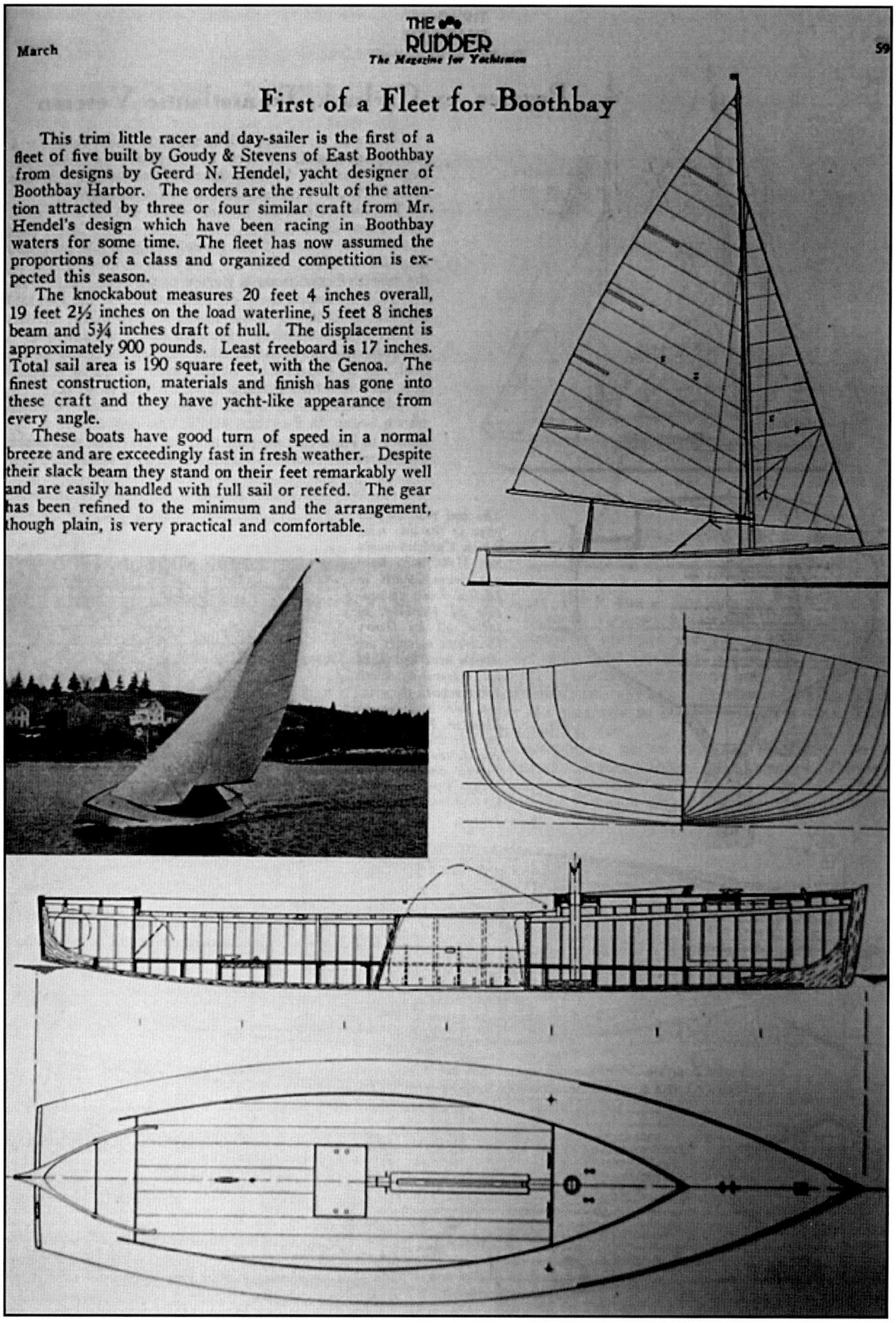

March — THE RUDDER — The Magazine for Yachtsmen — 59

First of a Fleet for Boothbay

This trim little racer and day-sailer is the first of a fleet of five built by Goudy & Stevens of East Boothbay from designs by Geerd N. Hendel, yacht designer of Boothbay Harbor. The orders are the result of the attention attracted by three or four similar craft from Mr. Hendel's design which have been racing in Boothbay waters for some time. The fleet has now assumed the proportions of a class and organized competition is expected this season.

The knockabout measures 20 feet 4 inches overall, 19 feet 2½ inches on the load waterline, 5 feet 8 inches beam and 5¾ inches draft of hull. The displacement is approximately 900 pounds. Least freeboard is 17 inches. Total sail area is 190 square feet, with the Genoa. The finest construction, materials and finish has gone into these craft and they have yacht-like appearance from every angle.

These boats have good turn of speed in a normal breeze and are exceedingly fast in fresh weather. Despite their slack beam they stand on their feet remarkably well and are easily handled with full sail or reefed. The gear has been refined to the minimum and the arrangement, though plain, is very practical and comfortable.

The March 1935 issue of The Rudder *contained an article titled "First of a Fleet for Boothbay" that described significant movement among Boothbay Harbor sailors toward the development of a Boothbay 20' one-design racing class based on the Hendel-designed centerboard sloop shown here.*

The Rudder magazine.

Additional evidence of the area's growing interest in developing a one-design fleet comes from a front-page article in the March 8, 1935, edition of *The Boothbay Register*, which reported on the "Brutal Beasts" built by Norman Hodgdon Jr. "With close to seven boats owned at Boothbay Harbor, and good prospects that their number will increase before long, it seems likely that the Boothbay Harbor Yacht Club is to sponsor a series of races for this one-design class during the coming summer."

Maynard Bray wrote in "The Boothbay Harbor One-Design," *Small Boats* 2007, "The mid-1930s were the bleak Depression years when small boats rather than big ones were receiving attention—quality attention—from Boothbay region designers, builders, and sailors. Boats with long waterlines and short overhangs began dominating the Boothbay racing fleet in those days, and top echelon designers took notice. A long waterline means a faster boat; boats of these proportions came not only from Hendel and Burgess, but from Charles Hodgdon of East Boothbay's Hodgdon Bros. yard, and from L. Francis Herreshoff."

Over the period of 1934–38, while he was working with Burgess on the *Ranger* project, Hendel's racing sloop design work was ongoing, and he became highly regarded for his successful 18' and 20' centerboarders. He was then a resident of Boothbay Harbor, joining the Boothbay Harbor Yacht Club (BHYC) as a member about 1938 and serving on both the Regatta Committee as Measurer and the Entertainment Committee.

An article written by Pete Welles Jr. in the *Boothbay Register* of August 22, 1952, contained specific detail about the local development history of the BHOD:

> During the latter years of the famous motorboat races at Boothbay Harbor, a small group of sailing enthusiasts embarked on developing sailboat racing for the region. The first race year was 1934, and by 1935 the Cape Cod knockabout and the handicap class were well established. Knockabouts were raced as a class with other dissimilar boats making up the handicap group. It was from these handicap races that a need appeared for a standard type boat that would be bigger and faster than the small knockabout.
>
> Hendel, who was working for Starling Burgess as a yacht designer, became interested in developing a one-design class for Boothbay Harbor and Christmas Cove (across the Damariscotta River in the nearby town of South Bristol). L. Francis Herreshoff, who was visiting Burgess, also became interested in the one-design possibilities.
>
> With the help of Herreshoff and Burgess (in 1934), Hendel drew up the plans for a 20' centerboard sloop. Four of these boats were built and were bought by local yachtsmen George Foster, Frank Spencer, and two by Philip Haselton Sr. These four boats (*Red Spot, Medrick, Leeward, and Windward,* all built at Goudy & Stevens) were raced in the handicap class, but even after being changed from centerboards to keels with spade rudders (weighing 900 lbs. and installed by Goudy & Stevens), they proved too slow.

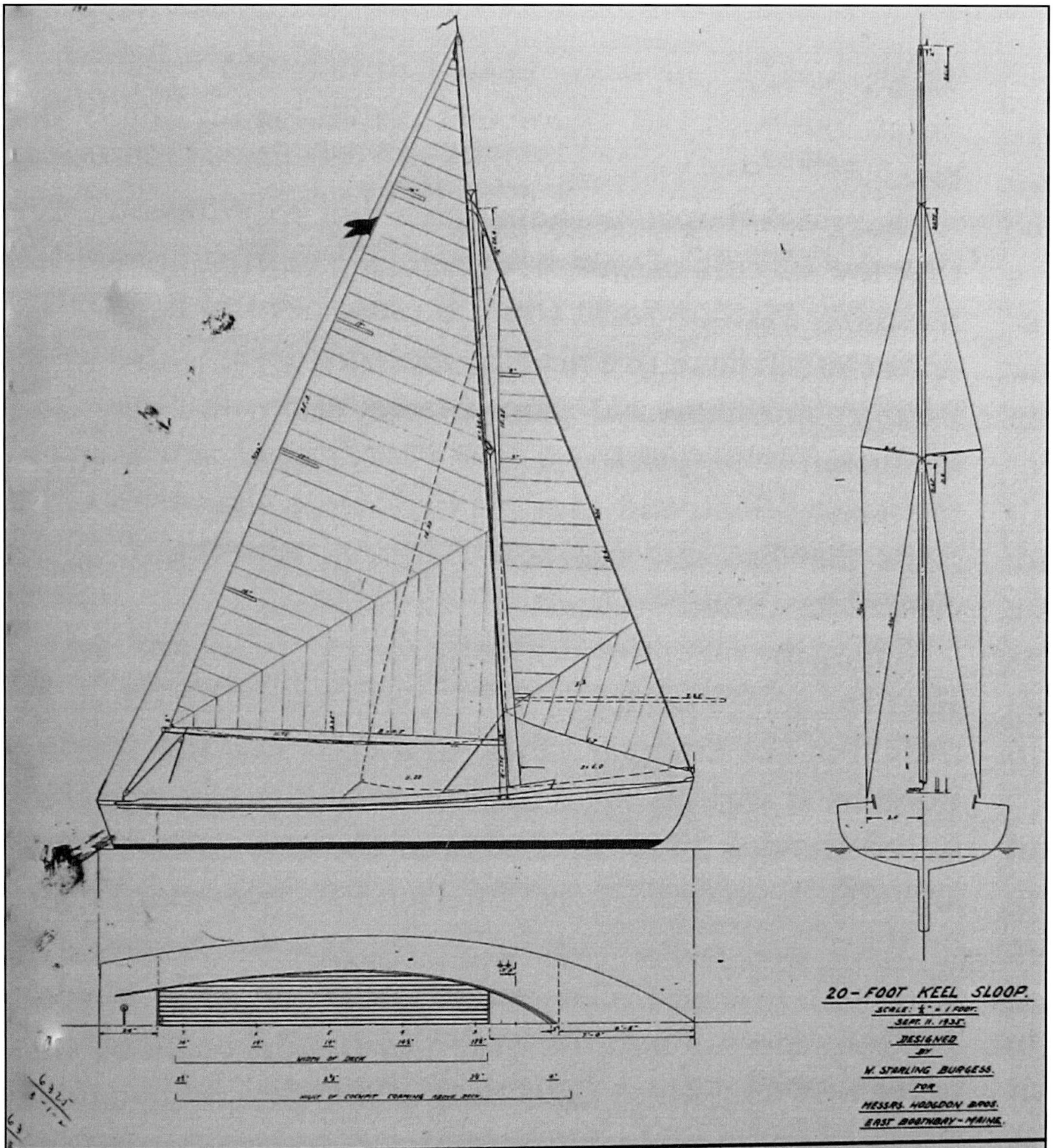

Sail plan for 20-foot keel sloop *Bunny*, designed for Hodgdon Bros., August 1935. W. Starling Burgess.

© Mystic Seaport, Daniel S. Gregory Ships Plans Library, Burgess-Wolff Coll., 43.76.4.

Bunny II.

Sanderling.

Waterline, newsletter of the Boothbay Harbor One Design Association.

In 1935, W. Starling Burgess "designed a 20-foot daysailer which was to compete in the popular Boothbay Harbor 20-foot open design class that Geerd Hendel had been promoting. Short-ended, firm-bilged with a generous modern sail plan, the keel sloop was built for Burgess by Hodgdon Brothers at a cost of $450, with the understanding that Burgess could sell her back to Charles Hodgdon for $300 'if delivered at [the Hodgdon] yard in good condition any time before May 1, 1936.' Named *Bunny* (*Bunny II*), which was Burgess's pet name for 'Nannie' [his fourth wife, Anna Hope Dale, who married Edward Biddle in 1922 and divorced in 1933, and whom Burgess married in 1933] and was also the name of a previous Boothbay Harbor 20 [*Bunny I*] that had provoked Burgess's interest in the class, the boat was launched and commissioned at East Boothbay on a Saturday morning in early October."[42]

Burgess's 1935 *Bunny II* sailed well, and Burgess hoped the BHYC would be convinced to move beyond its 20' open class in favor of a new 20' one-design class based on *Bunny*'s lines. This Burgess-designed/Charles Hodgdon-built racing sloop was received with interest by the BHYC committee, but it was later rejected because Hodgdon Brothers wished to be the exclusive builder of this class. The BHYC committee thought that anyone should be able to build them.[43]

In correspondence dated November 6, 1935, Burgess said that to build *Bunny II*, "The price complete including the loose-footed Ratsey Mainsail and jib is $675." Burgess further gave, in a November 12, 1935, letter to the editor of *Yachting* magazine, his opinion that the boat "has proved so fast as to completely outclass the Hendel boats in nine informal races."[44]

In his 1988 oral interview, Hendel provided his opinion that Burgess's 20' keel boat *Bunny II* was lightly ballasted and was very "tender" on windy days.

Llewellyn Howland III in *No Ordinary Being* noted that W. Starling Burgess had designed several early 20-footers that sailed in Boothbay waters, including *Bunny II* in 1935, which was very similar to the design that eventually became the BHOD. Burgess designed *Bunny II* with a loose-footed mainsail, however, which was not a success.[45]

Then in 1936, Hendel designed *Sanderling*, a Boothbay 20–class keel sloop to which he made modifications to the keel designs of his first boats. *Sanderling* was built by Norman Hodgdon Jr., and she represented a significant evolution of Hendel's design work. *Sanderling* was fast, but she was thought to be too expensive at about $650. Hendel used her as his personal boat until the next year when his *Loon* was built.

Sanderling was purchased in 1937 from Hendel by Oliver Garceau, a Harvard professor who raced her in the handicap class.

Russell Burgess moved the vessel to the Stonington, Connecticut, area in the 1960s, and she was sold to a Mr. Miller from Stonington in the mid-1970s. Efforts of the Boothbay Harbor One Design Association (BHODA) since 2005 to locate *Sanderling* have been unsuccessful.[46]

Sanderling *in 1969.*

Russell Burgess photo, Hendel Photograph Collection, Maine Maritime Museum.

At the August 1937 annual meeting of the BHYC, "a great deal of interest was shown in the development of a one-design class for Boothbay Harbor to supplement and perhaps ultimately take the place of the handicap races, which had been the major yacht club activity during the last few years. Vice Commodore Alden Reed commented, 'Next summer we're going to have a One-Design Race, so if you want to race, you'd better get one.' A Regatta Committee consisting of Oliver Garceau, Philip Haselton, Geerd Hendel, Eugene Tavenner, and William Leach was chosen and instructed to work toward this end."[47]

In October 1937, Hendel designed the 21' keel sloop *Loon* and had her built at his own expense by Norman Hodgdon Jr. She was proposed as a one-design class boat but was initially rejected by the BYHC design committee as being too slow.[48]

The summer 2005 issue of *Waterline*, the newsletter of the BHODA (Vol. 1, Issue 2, pgs. 1 and 5) contains an expansive narrative compiled by newsletter editor Alden Reed II about *Loon*. In summary, *Loon* was launched in November 1937, and sea trials were conducted over several weeks, with Jim Stevens of Goudy & Stevens sailing an earlier Hendel-designed Boothbay 20 class built in 1934 by Goudy & Stevens (either *Windward* or *Leeward*) as a boat against which to test *Loon*'s performance. Hendel was disappointed with *Loon*'s initial trials and made modifications to make her faster, but they added to the construction costs. Alden Reed II observed that modifications perhaps included changes to *Loon*'s keel shape and replacement of her cast-iron shoe with a lead shoe to make the boat sail more upright in stiff winds. These changes enabled *Loon* to increase her speed.

Loon *at the old in-town location of the BHYC.*

Norman H. Hodgdon Jr. photo, Alice Hodgdon Thompson Collection.

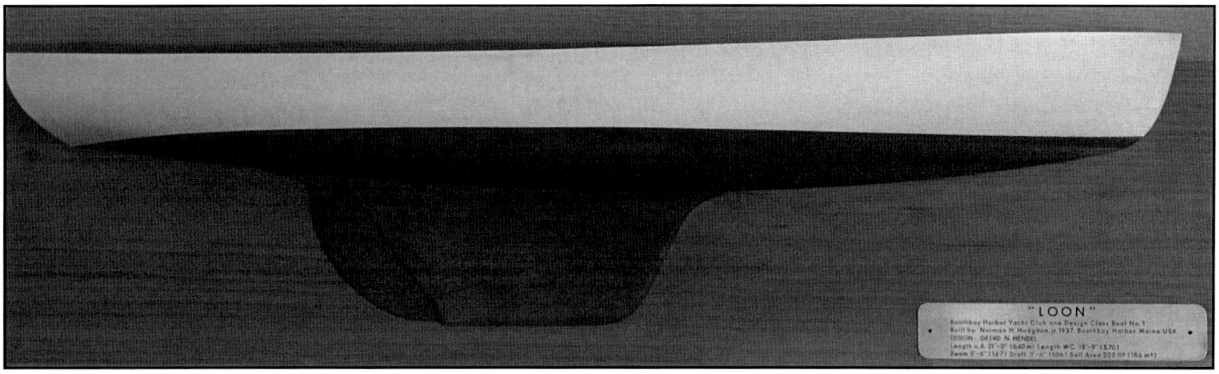

Loon *half-hull model.*

Barthold and Ingrid Brümmer Collection.

The Loon *half-model's plaque.*

Barthold and Ingrid Brümmer Collection.

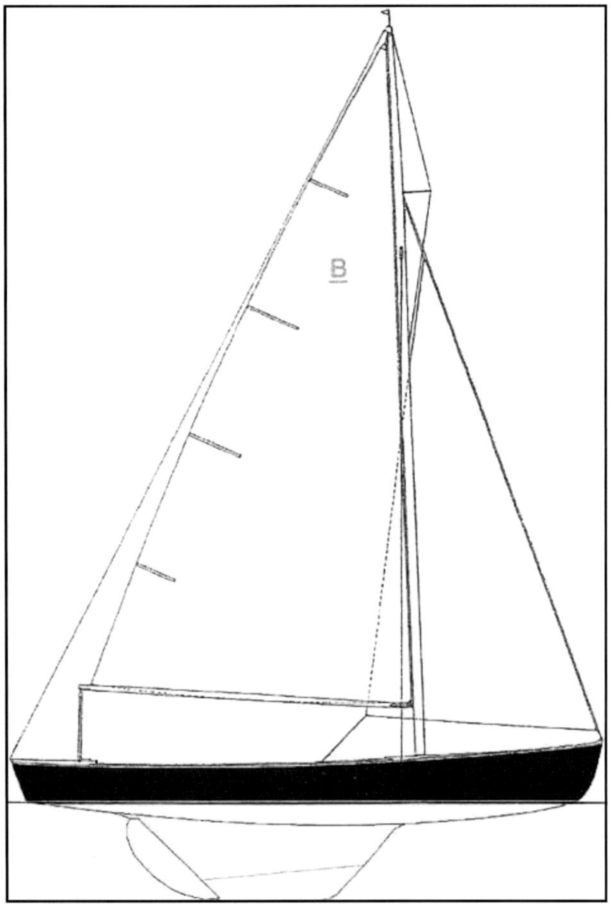

A 21' racing sloop design by Geerd N. Hendel.
Length: 21'0" (6.4 m)
Waterline: 18'9" (5.72 m)
Beam: 5'6" (1.71 m)
Displacement: 1950 lbs. (884.505 kg)
Draft (fin keel): 3'6" (106.68 cm)
Mast height: 28'3.5" (8.62 m)
Sail area (main & jib): 200 sq. ft. (18.5806 m²)
Lead ballast: 900 lbs. (408.23 kg)
Crew (racing): 2–3

Waterline newsletter, December 2007.

By 1938, the need for a one-design class became even greater because people were winning BHYC races based on the cost of their boats rather than because of a sailor's ability. A committee was formed and decided to have designs submitted for a racing sloop that could be constructed for about $650.

Arthur Stevens of the Goudy & Stevens boatyard and Hendel had become good friends, and Stevens worked with Hendel to develop options for the BHYC one-design competition, including building towing models for tank testing.

On October 1, 1938, Geerd Hendel finished and submitted his one-design proposal to the BHYC's one-design committee. Later in the month, he was awarded the design commission for the new 21' keel racing sloop, which was originally called the Boothbay Harbor Yacht Club One Design. For Hendel, then, the BHOD represented the culmination of nearly a decade's work of designing, building, and testing his ideas for fast racing sloops.

As to the BHOD award to Hendel and not Burgess, Llewellyn Howland III observed "That the design of the new 21-footer had arguably been influenced by Hendel's important work on the J boat *Ranger* and other Burgess projects was an irony not, one suspects, wholly lost on Burgess himself."[49]

The man who provided the initiative to get the first of the BHOD's constructed was Alden Reed, who owned the first two boats and financed others.

The first BHOD boat, *Fraülein* (owned by Alden Reed), was built by Benjamin "Benny" Rand in Boothbay Harbor, and the second, *Mädchen* (also owned by Alden Reed; the name meant *girl* in German; later renamed *Poppy* and then *Otter*) was built by Rand and Norman Hodgdon. The first two boats were given German names in tribute the Hendel's German heritage.

76 YACHTING

The Boothbay Harbor Y. C. One-Design Class

THESE interesting boats might be considered the development of the open class of 20-footers which has been racing at Boothbay since 1937. Those boats came from the boards of several designers and were built in several yards. There were owners, however, who preferred a one-design class so that they could have craft which would not be outbuilt.

They are from the board of Geerd M. Hendel, of Boothbay Harbor, and are of the following dimensions: Length over all, 21'0"; length on water line, 18'9"; beam, 5'6"; draft, 3'6". Displacement is about 1750 pounds. The sail area (loose-footed mainsail and overlapping jib) is 200 square feet. A medium sized spinnaker is used.

Construction is to rigid specifications, with oak frames, ½" cedar planking, white pine deck, canvas covered. Coaming and trim are of mahogany. The fin is of yellow pine with lead keel. The boats are bronze fastened throughout. Spars are of Sitka spruce, the mast of hollow box form. Single shrouds are used with turnbuckles below deck. Cockpits may be shaped to suit the owner's requirements, within certain limits. Deck fittings and rigging details may be varied as owners desire.

The class was started in 1938 with six boats and the fifteenth is now under construction. Most of them have been built by Norman H. Hodgdon, Jr., Boothbay Harbor, and Goudy & Stevens, East Boothbay, and two by Wesley Mahan, Pemaquid Harbor, all in Maine. The boats may be built in any yard the owner selects but, upon completion, are checked by a member of the club's race committee. They trim well, with or without crew, and are well balanced and fast.

The sail plan shows a modern rig with a moderate spread of canvas

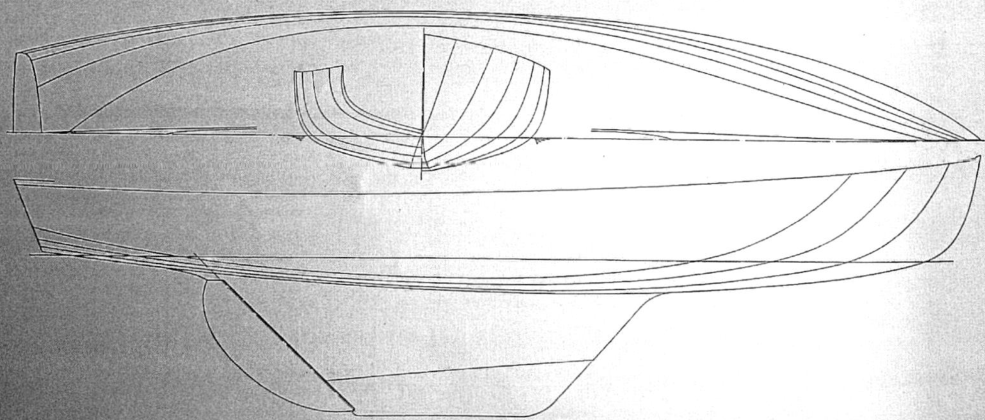

The new boats are of the fin keel model with long and easy fore and aft lines

Yachting magazine, June 1939.

Hull number three, *Hardalee* (owned by Philip H. Haselton), was built by Goudy & Stevens during the winter of 1938–39. A total of ten boats were built and raced the following summer. In addition to the first three, the other seven boats were *Dolly* (owned by Alden Reed and built by Rand and Norman Hodgdon), *Lea* (owned by Edmund B. Spaeth), *Eight Ball* (owned by Eugene Tavenner), *Psyche* (owned by Melville H. Ireland), *Panda* (owned by Ireland), *Harvey* (owned by MacFall), and *Windward* (owned by Haselton). *Hardalee,* sailed by thirteen-year-old Phil Haselton, was the first season winner, overcoming such competitors as Starling Burgess and Geerd Hendel. The following winter, Norman Hodgdon built *Caper, Eight Bells,* and the *Bluebird;* Goudy & Stevens built *Linda, Jinx Two,* and *Gaviota;* while Ben Rand built *Hit-Or-Mis* in 1940. *Old-Cro, Pilgrim,* and *Sea Rascal* were added in 1940–41. Twenty boats raced in the 1941 season.[50]

The *Boothbay Register* of October 7, 1938, reported, "This new class boat was designed and is well adapted for Boothbay Harbor and surrounding waters, is fast and able, well-proportioned and balanced, and a pleasure to sail. Two of the [four] new sloops are to be built by Norman Hodgdon Jr. of this town, and two [one of which was *Panda*] by Wesley Mahan of Pemaquid harbor."

With Hendel's one-design submission not having been approved by the BHYC until October 1938, it's presumed that the boats racing in the summer of 1938 were likely *Bunny II, Sanderling, Red Spot, Medrick,* and *Leeward,* which were all Boothbay 20s, and possibly Hodgdon's 1935 20' *Popeye* and *Mink. Loon,* of course, was 21'.

Hendel and Starling Burgess continued to actively race in the BHOD's in the Boothbay area during the years leading up to World War II.

With the adoption of the 21' BHOD, the BHYC's open-class 20-footers were placed in a handicap racing class. *Loon* was included in the handicap class, and Hendel continued to race her into the late 1930s and early 1940s.

In his 1988 interview, Hendel acknowledged his appreciation of the three boatyards in Boothbay/East Boothbay (Hodgdon Brothers, Rice Brothers, and Goudy & Stevens), which he admired for the many skilled carpenters in their employ who built quality boats.[51]

L. Francis Herreshoff, a frequent visitor to the Boothbay area, was likely inspired by the designs of his friends Burgess and Hodgdon when he drew *Koala,* a 22'6" sloop, in 1937. She showed a striking resemblance to the early Boothbay daysailers.[52]

Jim Stevens of Goudy & Stevens and his wife, Evie, moved to Camden in the early 1940s, when he was hired as lofting manager at Camden Shipbuilding, which was then constructing vessels for WWII. Hendel became employed there in 1941.

During that year, Hendel moved *Loon* to Camden harbor, and he and Stevens sailed her in Penobscot Bay waters. Stevens joined the U.S. Navy in WWII and returned to Boothbay Harbor thereafter, where he bought out Wallace Goudy and formed the Goudy & Stevens boatyard. Jim was also a founder of the Boothbay Region Historical Society.

Hendel sold *Loon* in 1945, and she was then sailed out of Rockland, Pemaquid, Manchester, and Boothbay harbors under various owners. She was sheathed with fiberglass for a time, but that was removed by boatbuilder David Nutt at Southport, Maine, under Hendel's guidance. Her history during the late years of the twentieth century is unknown at this point, but she was said to have been spotted lying in a mud bank in South Portland. In 2001, Roger Gould of Gloucester, Massachusetts, purchased *Loon* in Portland and had her restored.[53]

The restoration of *Loon*, now owned by Allan Miller and Pam Burke, was carefully completed by Allan himself in 2018.

Following are four commentaries describing the much-respected and admired Boothbay Harbor One Design.

» The website for Artisan Boatworks, Rockport, Maine (www.artisanboatworks.com) offers this perspective on Hendel's Boothbay Harbor One Design:

> Designed by Geerd Hendel, one of Maine's best designers and unfortunately one of our least-known, the Boothbay Harbor One Design is something quite different from many of the boats we are describing on this site.
>
> While she is not a new design, in some ways she is more like a modern 'Spirit of Tradition' racer than what we usually think of as a classic One Design.
>
> Unlike many classics, she has very short overhangs, giving her a long waterline for a high potential top speed. Due to her light displacement and her very long, relatively straight hull lines, she is capable of sailing at her top speed a good deal of the time. Because of her light displacement, she does not have the dramatic curves of other designs—her shapes are more subtle, but they are sophisticated nonetheless, and with familiarity will capture the eye as much as any boat.
>
> At the same time, she is a simple and roomy boat, uncluttered, and easy to sail. While it may not be immediately evident, she is also a comfortable boat, with a choice of sitting on the wide side decks outboard of the cockpit or down on the cockpit sole, where the coaming comes at good backrest height, and one's feet can brace comfortably on the lee side. While the cockpit sole may not constitute a "seat" per se, it is actually a great place from which to sail, as one is seated well down in the boat, safe, relatively dry, and with good visibility all around. She has excellent "feel" under sail.
>
> With a boom tent, one could sleep on her cockpit sole with the anticipation of a fast and exciting passage to the next anchorage the following day.[54, 55]

» In an article in *The Working Waterfront*, July 1, 2007, Roger S. Duncan writes the following observations on the BHOD:

New blood is flowing for an old class of boats. David Nutt is building two new Boothbay Harbor One Designs at his shop in Edgecomb, Maine. These lovely boats were designed by Geerd Hendel in 1937 for the Boothbay Harbor Yacht Club. The Club wanted a family boat, stiff enough to stand up to a summer northwester, easy enough to handle for the learners, but fun for adults and fast enough to race. A number were built over the years by Norman Hodgdon, Ervin Jones, and George I. "Sonny" Hodgdon. To keep up with the times, a number of fiberglass One Designs were built, but this ended when the molds were lost in a fire 27 years ago. Over the years, some boats sat in barns unused, and others were not raced and neglected. Seven or eight were still racing last summer.

The Boothbay Harbor One Design Association has been promoting the class, and the time is right to reintroduce the design. In the fall of 2006, David Nutt made agreements to build boats for two new customers. To be competitive, fairly, the new boats must be exactly the same shape and weight as the originals. Robert Stephens at Brooklin Boat Yard in Brooklin [Maine] made patterns by computer on sheets of Mylar from Hendel's original lines to eliminate possible errors in lofting. This also assures a buyer, regardless of who builds the boat, that each will be of precisely the same design. He worked with Nutt's plans for construction to be sure the weights would be the same, even to specifying the number of coats of finish paint.

On January 4, 2007, Nutt started getting out and setting up the molds. The boats were built upside down without keel or frames. Floor timbers cut to match with the mold were dropped in and each temporarily tacked to its adjacent mold. Three bulkheads were set in the same way. The boat was then planked with half-inch Alaskan yellow cedar, each edge-glued to its neighbor and to floors and bulkheads. A layer of fiberglass saturated with epoxy resin was then stretched over the whole structure, and the boat, complete with floors and bulkheads, was lifted off the molds. There is no interior wooden stem. Where the planks come together at the bow, they are glassed together inside and out.

Short decks fore and aft are of plywood on fir frames covered with fiberglass and epoxy. The keels are made separately of African mahogany, ballasted with 900 pounds of lead and bolted up through lead, wood, planking, and floors. Finished, each boat will be 21' overall with a beam of 5'6" and draft of 3'6".

One of the boats will have a wooden mast and boom, and the other will have aluminum spars, each carefully adjusted to the specified weights. Sails will be made by Maine Sailing Partners in Yarmouth, both boats with roller furling jibs. A new boat will cost little more than rebuilding a neglected one.

When these two new boats face the starting line this summer, each will be exactly the same shape and weight as specified for the original Boothbay Harbor One Designs.[56]

Boothbay Harbor One Designs race in this photo from the late 1930s or early 1940s. Sail #2 is probably Mädchen *(owned by Alden Reed); later named* Poppy *and then* Otter. *Sail #5 is* Eight Ball, *now* Tia.
Geerd Hendel Collection, courtesy of his stepdaughter, Brigitte M. Gefrich.

Alden Reed II commented as follows concerning this photograph: "If the picture is from 1938, then #2 would be *Mädchen* owned by my grandfather. He had a number of BHOD's built to help build up the class, because prospective owners would buy a boat in the water but were less likely to buy a boat if they had to order one and wait for it to be built. It gets confusing, as he sold #2 when it was new but retained the sail number, so it ended up on *Dolly*, an experimental BHOD built in 1940. *Dolly* was constructed out of spruce because it was the lightest wood and therefore, in theory, made a faster boat. At that point the BHYC passed a rule banning the switching of sail numbers."

» *The Wiscasset Newspaper* reporter Debbie Swan described a meeting of the BHODA (while she was serving as president) at Juniper Point on August 24, 2013, which was posted online Friday, August 23, 2013. At the meeting, boatbuilder David Nutt discussed building new One Designs and Swan reported as follows:

The Boothbay Harbor One Design is one of the nicest 21-foot classic sloops ever designed. This is a boat loved by everyone who has owned one. With their heavy keel, they are a very safe family boat. None have ever capsized. The fleet has come and gone over the years, and we appear to have a resurgence of interest in the boat. Although there have been 56 built, if someone wants to buy one, they are hard to find.

David Nutt believes it is time to once again build Boothbay Harbor One Designs in fiberglass. He plans to build a mold and produce a fiberglass vacuum-bagged cored hull and deck, a fiberglass hull liner and rudder. The boat will be finished with a choice of teak or mahogany trim. The hardware can be either traditional bronze or modern stainless and aluminum. The mast can be either aluminum or Sitka spruce. These boats will be ready for delivery in June 2014.

Boothbay Harbor One Designs have been built using four different construction methods over the years. The original boats were plank-on-frame classic wooden construction. Fiberglass boats were built in the 1970s. Recently Brooklin Boatyard (for the late Ted Leonard) and Maynard Bray at the WoodenBoat School each built one using an epoxy cold-molded construction technique. During that same year, David Nutt built two using a strip-built fiberglass hull using epoxy resin systems (*#1 Nutt Job* and *C-Fever*). David still has the mold for this construction technique and can build to order. There are a number of the original classically built wooden One Designs sailing in the area. As wonderful as these boats are, they are a relatively high-maintenance boat and require a considerable level of skill to keep in top shape. Several of the classic plank-on-frame One Designs have been resurrected by laminating two 1/8" veneers to the exterior of the hull with epoxy resins and then fiberglassed on the exterior. This gives the boat greater strength and rigidity than the originals ever had and minimizes the maintenance issues on the bottom and topsides. This resurrection technique can be done to any existing wooden hull.[57]

» Maynard Bray, author and regular contributor to *WoodenBoat,* who serves as the magazine's technical editor, says of the BHOD:

It's a thrill to sail a BHOD either singlehandedly or with a friend or two. You tuck a cushion under your fanny and sprawl on the floorboards, where you're out of the wind and much of the flying spray, yet you can see all around since your line of sight is under the boom and not obstructed by it or the mainsail. In tacking, the mainsail takes care of itself. Although the jib has to be let go and resheeted when changing tacks, it's a small sail that blows across to the new lee side and sheets in easily once there—hardly needing the winch except in strong winds. With half her total weight at the bottom of her keel, the BHOD can be driven harder than a centerboarder of equal size. A reef is seldom needed, and capsizing is almost unheard of.[58]

Wooden hull production ended in 1966 with Tom Tavenner's *Galatea* (sail #37), and the first fiberglass BHODs were introduced in 1973 molded by Coastal Plastics and completed by Sonny Hodgdon.[59]

Hodgdon 21/Christmas Cove One Design/ Hodgdon Christmas Cove 21/and Others

A class of 21' sloops, called the "Hodgdon 21," were developed by Charles Hodgdon and six were built in the mid-1930s.

The first two Hodgdon-built sloops with a fin keel (1935) were reportedly *Popeye*, which was 20'4" LOA with a 5' beam (6" narrower than the BHOD), had a sail area of 200 square feet, a high bow, and a flat transom, and *Mink*. Their built-down "wineglass" bilge with a fin keel is considerably different from the relatively flat bottom and fin keel on the Burgess-designed *Bunny II* and the Hendel-designed Boothbay 20s and BHOD. (In the summer 2006 edition of *Waterline*, Volume 2, Issue 2, editor Alden Reed noted that *Popeye* and *Mink* were often referred to as early Christmas Cove One Designs.)

Mink's *"wineglass" bilge shape and flat transom on display at the Maine Maritime Museum.*
Roger Allen Moody Collection.

Christmas Cove One Design

In 1938, Charles Hodgdon built five sloops, known locally as the Christmas Cove One Design (CCOD), which, unlike the 1935 design, had radiused transoms.

The Hodgdon 21 Class One Design (HB-21 or H-21), a redesign by George I. "Sonny" Hodgdon of his uncle Charles Hodgdon's 1938 racing sloop, was introduced into the Christmas Cove Yacht Club races in 1958 and proved very popular.[60]

In 1971, the fiberglass Christmas Cove 21 Class (based on the Hodgdon 21 *Goose* per Richard "Rick" Prescott Jr.) and incorporating several design modifications suggested by Gus McDaniel, became the CCYC one-design racing sloop; today it's called the Christmas Cove One Design.[61]

WATERLINE **SUMMER 2006**

1938 Christmas Cove One Design Class

LOA: **20'4"**
LWL: **- -**
Beam: **6'**
Draft: **Keel 3'3"** **(Hendel estimated in letter Nov 2, 1938)**
S.A.: **200 sq. ft.**
Displacement: - -

In 1938 Charles Hodgdon built five keel sloops known locally as the Christmas Cove One Design (CCOD) Class according to late Rick Prescott. The transom on the CCOD is radiused unlike earlier Charles Hodgdon designs. Ballast is thought to be 900 pounds according to Sandy Holloway owner of *Gleam* (CCOD #1), who with Rick Prescott restored the boat in 2003. *Noel* is owned by Robert Nelson and restoration only required deck replacement (red mahogany) and refastening. *Hi Ho (CCOD #5)*, originally owned by Charles Prescott, was restored by Rick Prescott, according to current owner John Mitchell of Christmas Cove. *Calypso* is owned by Eddy Blouin of Riverside Boat Co. Newcastle, ME, sail number is unknown and boat needs restoration.

(Editor's note: Pete Welles's two part article sketching out the early history of the Boothbay Harbor One Design Class in August 1952 *Boothbay Registers* suggests W. Starling Burgess designed the Christmas Cove One Design. Although that is a possibility, the weight of evidence points to Charles E. Hodgdon).

Photo by Rick Prescott

Gleam Relaunching 2003

Above, Christmas Cove One Design, Gleam #1

Waterline, the newsletter for sailing enthusiasts in the Boothbay Harbor One Design and Sister Class Racing Sloops—Boothbay 20, Christmas Cove One Design, Great Lakes 21 (I-21), Hodgdon 21, and Christmas Cove 21.

Howland asserts, "The influences of both Burgess and Hendel could be seen in the new Christmas Cove class of one design 20-footers designed and built by Hodgdon brothers."[62]

Great Lakes 21 (GL 21)

In 1935, a sailing enthusiast named Harold Kneen from Cleveland, Ohio, vacationed in Christmas Cove, Maine. While in the area, he became an instant admirer of the local sloops, realizing they would be ideal for sailing in his home waters of Lake Erie. Soon after, he bought *Gigolo*, designed by Hodgdon Brothers; *Bunny* (*I*) designed by W. Starling Burgess; and *Valkyrie*, assumed to be a Burgess design, and had them shipped to the Mentor Harbor Yachting Club (MHYC) near Cleveland.[63]

To meet the MHYC's demand for more sailboats, Charles Hodgdon then designed a 21' sloop branded as the Great Lakes 21 (GL-21). Now renamed the International 21 (I-21), this class continues to be manufactured in fiberglass under the sponsorship of the International 21 Yacht Racing Association.[64]

Small Point One Design

The 19½' Small Point One Design class was W. Starling Burgess Design No. 80, which Burgess had prepared for Sumner Sewell and others at Bath, Maine, in late 1937.

Sewell, a leading figure in the summer colony at Casco Bay's Small Point, had been tasked with selecting the designer for a Small Point One Design class. He and Starling Burgess were aviation pioneers who shared many interests and had similar tastes in small craft.

"The result was a short-ended and very lively 19½-foot sloop with a cast-iron keel—a boat with an obvious family resemblance to Burgess's earlier design of the open-class Boothbay Harbor 20 *Bunny (II)* and his subsequent design of the Yankee One-Design, not to mention competing sloops from the boards of Geerd Hendel and Charlie Hodgdon....Neither Burgess nor draftsman Geerd Hendel had any say (or perhaps any great interest) in how the boats were to be built, since they had no financial stake in the class's long-term success."[65]

A Small Point Association report from the early 1950s recommended that any new boat be inspected by a naval architect for conformity with Geerd Hendel's 1949 drawings for the class.

The *Small Point Class Notes* fall 2006 issue noted under a paragraph titled "Class uniformity and new boats" that keeping the class healthy and competitive had been a subject of discussion since its founding.

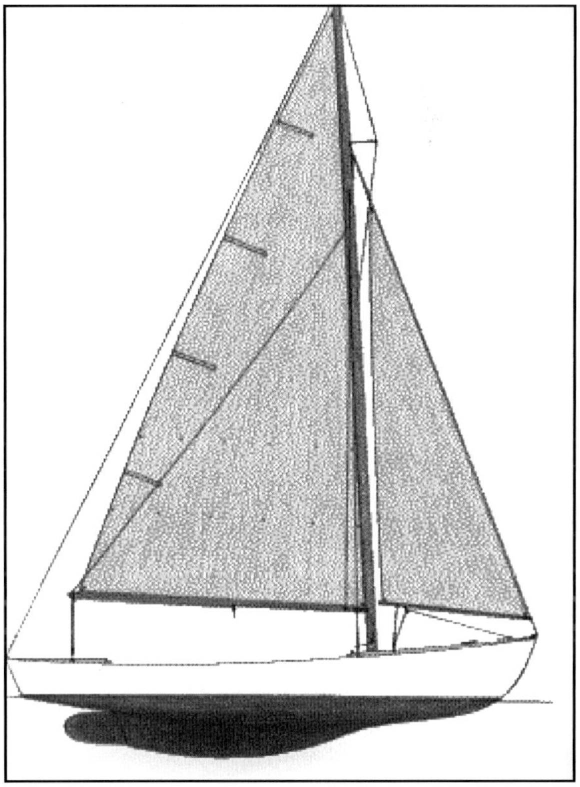

Small Point One Design.
www.SailBoatData.com.

The Yankee One Design (YOD) Class

W. Starling Burgess had just finished the design for his third America's Cup champion *Ranger* in 1937 when his design submittal won the New York City Yacht Club's Yankee One Design (YOD) Class design contest. This contest was to create the ultimate sailboat for yacht clubs' fleet racing. The impact of the Great Depression was severe, and New England sailors wanted to give New England boatyards the work of building their yacht club racing fleets.

The sailors behind the YOD class had been dismayed that a recently formed racing class, the 33' International One Design (IOD), had rules stipulating that its hulls had to be built in Norway. The first fleet of 25 Norwegian-built IODs were delivered to the NYC Yacht Club in December of 1936, and the YOD Association announced its design contest soon after.

The YOD design contest was promoted with a flurry of media coverage, and three prestigious America's Cup yacht designers were recruited as judges: W. Starling Burgess, L. Francis Herreshoff, and Frank Paine.

Then, in a behind-the-scenes move, contest organizers also secretly hired one of the judges, Starling Burgess, to enter the contest. "Despite some expressions of objectivity and collegiality, the Design Committee of the Yankee One Design Association had evidently made a side deal with Burgess that would ensure his participation in the design process and defray the cost of his draftsman Geerd Hendel's work on the project, as well as giving Burgess a reasonable (albeit certainly not overgenerous) fee, presuming that his design was the one selected for building."[66]

During Hendel's taped 1988 interview archived at the Mystic Museum Research Library, his interviewers focused on the YOD design result, seeking to learn from Hendel whether or not he had a lead role in the innovative hull design. Hendel demurred, however, saying that the YOD design ideas came from Burgess, and his role had been to translate them into design drawings.[67]

Burgess's final edited blueprints for a 30'6" LOA sloop were accepted by the YOD committee in 1937 and sent to the boatbuilder before the contest deadline had even passed and the other entries had been turned in.

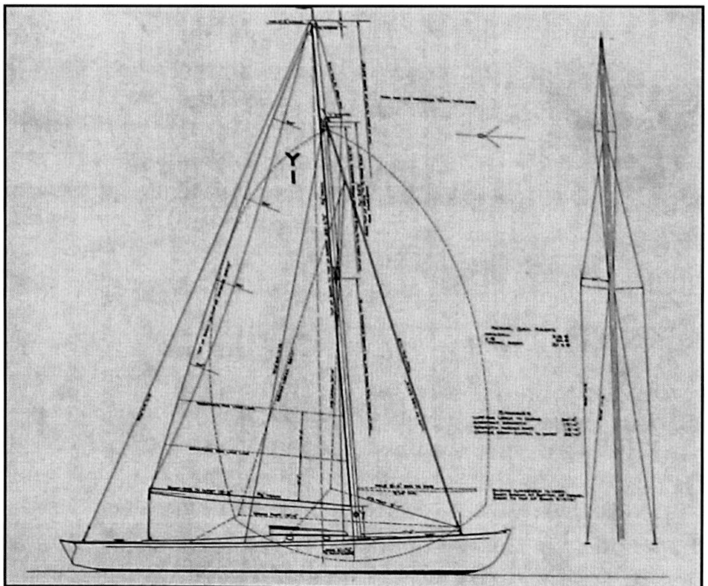

Yankee One-Design sloop, 1940. Sail plan.
© Mystic Seaport, Daniel S. Gregory Ships Plans Library, Miscellaneous Yachts Coll., YOD 4-4.

30'6" Yankee One Design sloop.
Hendel Photograph Collection, Box MS-310, Maine Maritime Museum.

"The YOD committee's design criteria required no formulas or 'Universal Rule' to be followed, which was unique for that era. ...Burgess designed a YOD with a relatively plumb bow, which was a milestone in sailboat design. Sailors were so used to seeing long graceful overhangs on racing boats that they incorrectly assumed overhangs made a boat go faster. But Burgess knew better. Forward overhang on a sailboat this size will slap against the chop. Burgess gave the Yankee an almost plumb bow, one that cuts through waves instead of trying to rise above them....The concave sides to the Yankee bow was another innovation. It not only helps the bow slice through the water, but the shape worked exceptionally well sideways, when the boat was heeling....No matter how fast it's going, when the YOD settles in and finds its power, it's typically heeling enough that the leeward side is actually acting as the bottom, and the long flat side with the concave bow serves as the bottom of a planing vessel, an extraordinary innovation for 1937."[68]

"The Yankee One Design was fast, able, distinctive in appearance, and a joy to sail...For whatever reasons, the boats never did find widespread support on Massachusetts Bay, much less on Long Island Sound, and only brief popularity Down East and on the Lakes."[69]

In his 1988 interview, Hendel also commented that the YOD was a pleasure to sail.

Yankee One Design Gemini, *built by the Northwest School of Wooden Boatbuilding, Port Hadlock, Washington.* Northwest School of Wooden Boatbuilding.

CHAPTER SIX

Aluminum-Alloy Sloop *Whistler*

From his mid-1930s work on *Ranger*'s aluminum-alloy masts and spars, Hendel became an early advocate of the use of aluminum in yacht-building. His work at BIW convinced him that aluminum alloy could withstand saltwater corrosion and would become useful as a boatbuilding material.[70]

In 1939, Hendel designed *Whistler,* the first aluminum-alloy pleasure craft ever built in the United States for saltwater use. It was an experimental yacht constructed with the cooperation of ALCOA and built at Rice Brothers in East Boothbay, Maine. The Rice Brothers yard had begun experimenting in 1936 with materials supplied by ALCOA.[71]

Rice Brothers Corporation boatyard, East Boothbay Harbor, Maine, September 1919.
Baxter M. Rice photo, Robert Rice Collection.

The Rice Brothers construction file describes *Whistler* as "The First Aluminum Yawl Built in the United States," so Hendel must have modified the original sail plan in favor of a sloop rig. She was 26' LOA with a beam of 9' and draft of 4'6".

Other construction notes say that the vessel was laid down on August 15, 1939, with a keel of yellow pine and a 2300-lb. lead shoe. The aluminum alloy used in the vessel's construction was donated by ALCOA. *Whistler*'s total displacement was 5200 lbs. Designed with accommodations for two, the main cabin area had a settee/berth on each side, and the galley area contained a stove, sink, ice chest, and dish racks. The head was forward on center, a stowage bin was located aft, and she had a 16-gallon water tank.

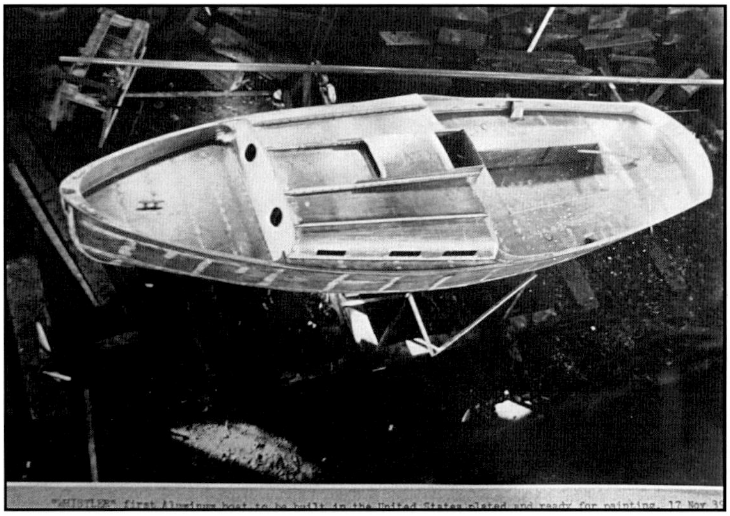

Alden Reed Collection, Boothbay Region Historical Society.

Port side of Whistler, *ground and ready for painting, November 17, 1939. All aluminum except the keel, which was yellow pine with a one-ton lead shoe.*

Geerd Hendel photo from the Robert Rice Collection.

In the May 2002 interviews with Boothbay Region Historical Society curator Barbara Rumsey, boatbuilder Jim Stevens (Goudy & Stevens), a good friend of Hendel's, confirmed that *Whistler* sank not long after being launched because a worker had placed the traditional copper penny under the mast when it was stepped. That caused electrolysis between the coin and the aluminum hull, resulting in a hole in her hull. *Whistler* was launched in 1940, and after repairs to her hull, Hendel owned and sailed her for more than 30 years.[72]

Folklore about the tradition of placing copper coins under masts in England and in Europe holds that such coins would enable the crew to cross to the afterlife if the ship were to sink. The coins were supposed to pay Charon, the mythical ferryman, for transporting the dead across the River Styx. Many other seafaring nations regarded the placement of coins beneath the mast as good luck for a ship. To avoid electrolysis, the coins are now often placed in corrosion-proof boxes before the mast is stepped.

Page 6

Former Iron Worker Designs Aluminum Sloop

The casual visitor to the Rice Brothers boat shop in Boothbay Harbor, sees what to him is a strange sight. For under construction in this shop is an all aluminum boat. Not a large passenger ship or a palatial yacht but a 26 ft. long auxiliary sloop. The craft was designed by and is being constructed for sale by Geerd Hendel, an ex-member of the drafting room crew at the Iron Works.

Work on the aluminum sloop was started in August, 1939, and it is expected that she will be launched some time this coming summer. The little craft, which can accommodate six passengers, will be entirely riveted with the exception of the gas and water tanks. Experiments were tried with welding, Mr. Hendel states, but proved so difficult that it was decided that he bes results would be obtained by riveting.

The sloop, which is being constructed with the cooperation of Jack Chesley of the development division of the Aluminum Company of America, is 22 feet, seven and one-half inches long at the water line, has a beam of seven feet, seven inches, a draft of four feet, six inches, a displacement of 4,750 pounds and a sail area of 300 square feet. The aluminum mast, which is now under construction in Pittsburg, Penn., will be 35 feet, six inches high. She will be powered by a 7/18 h. p. Red Wing "Meteor" engine. She has a hard pine keel with a lead shoe and a total of approximately 7,500, one-quarter inch rivets will be used in her construction.

Mr. Hendel was educated in Bremen, Germany, and came to this country to learn the English language but because the future abroad appeared so dark, he took out citizenship papers in this country. It is interesting to note, according to Mr. Hendel, that the price for aluminum boats will be from six to seven times higher per pound than steel, but beacuse there is more aluminum to the pound and because it is easier to work with than wood, it is believed they will be practical.

Mr. Hendel, a naval architect of merit, was the author of a paper which appeared in the March issue of "Yachting."

The Observer, *Bath, Maine, Vol. 1, No. 18, February 21, 1940, p. 6.*

Geerd Hendel Collection, Box MS-310, Maine Maritime Museum archives.

ALCOA Aluminum ad featuring Whistler.
MotorBoating magazine, February 1946, inside front cover.

The text in this January 1946 ALCOA ad reads:

> Made of Alcoa Aluminum—designed by Geerd Hendel, naval architect, and built in 1939 by the Rice Brothers Shipyard in East Boothbay, M[aine]...the *Whistler* is still ship-shape in every respect. A sleek, seaworthy craft for either fresh or saltwater sailing.
>
> Light weight and durability are main features of aluminum boats in every class. In addition, they are always tight, laid-up or afloat they will not rot, dry out, or shrink. Maintenance is greatly simplified...saving weekends for sailing fun.
>
> Alcoa Aluminum is a proud boatbuilding material. More and more pleasure craft builders are switching to aluminum construction. ALUMINUM COMPANY OF AMERICA, 1838 Gulf Building, Pittsburgh 19, Pennsylvania.

ALCOA ad featuring Whistler.

Yachting magazine, November 1945.

The text is difficult to discern in this November 1945 ALCOA ad, but reads as follows:

> "She's Tight as a Tick"
>
> Watertight, of course is what we mean. Because this sleek seaworthy 26' auxiliary sloop is built of Alcoa Aluminum Alloys...sheet, shapes, and railings...she'll never part at the seams or rot. In salt water or fresh, she'll take the whitecaps with speedy, graceful ease.
>
> All hull, deck, and cabin trunk parts are aluminum...all framing, stiffener, and rails are aluminum shapes, cleats, chocks, running lights, and port light frames are cast aluminum, tanks are welded aluminum sheet, and her mast is a single piece aluminum alloy extrusion tapered at the top.
>
> She was designed and built by Geerd N. Hendel of Boothbay Harbor, a man who knows his boats, and he knows the seaworthiness of aluminum marine alloys. Ask your builder or naval architect about the use of aluminum in that new boat for you.
>
> ALUMINUM COMPANY OF AMERICA, 1838 Gulf Building, Pittsburgh 19, Pennsylvania.

This ad was also placed in the December 1945 issue of *MotorBoating* magazine and likely appeared in other boating periodicals as well.

The Rice Brothers construction file in the Robert Rice Collection notes that *Whistler* was sold by Hendel for $4,000 in 1974 to William F. Stone of Bangor, Maine, and the Red Wing engine was replaced by a Universal Atomic 4 gasoline engine rated at 25-hp in 1977. Mr. Stone had listed the vessel for sale in 1979 for $5,000, but the trail of ownership has unfortunately disappeared for the 1979–2005 timeframe. Certainly, for some of the later years of that 26-year period, the boat was in storage at Mid-Coast Marine in Winterport, Maine.

In 2011, *Whistler* was listed for sale on the internet as a "project" boat by Mid-Coast Marine. That yard had acquired ownership for the nonpayment of storage fees. In 2018, the Mid-Coast Marine

Whistler *launched in East Boothbay at the Frank Rice boatyard in 1940.*

Rice Brothers, Box No. 2,
Boothbay Region Historical Society.

owner told the author that he had no records of the vessel, nor could he recall the name of the owner who had stored the vessel in the yard.

In a March 1, 2018, interview with the author, *Whistler* owner Seth Yentes said his family members have always loved sailing. In 2005, when he was a student at the University of Maine in Orono for about a year and a half, he and his sister were sharing an apartment and found it very costly. They began thinking they could buy a boat and put it on-the-hard in Orono near the university campus to live aboard. Their concept plan was to pay off the purchase price within a year and then moor it in Belfast harbor and have summertime fun sailing it on the coast. His stepfather was especially excited about steel-hulled boats and had heard of an aluminum-hulled sailboat at Mid-Coast Marine in nearby Winterport. They looked at *Whistler* and knew it wasn't big enough

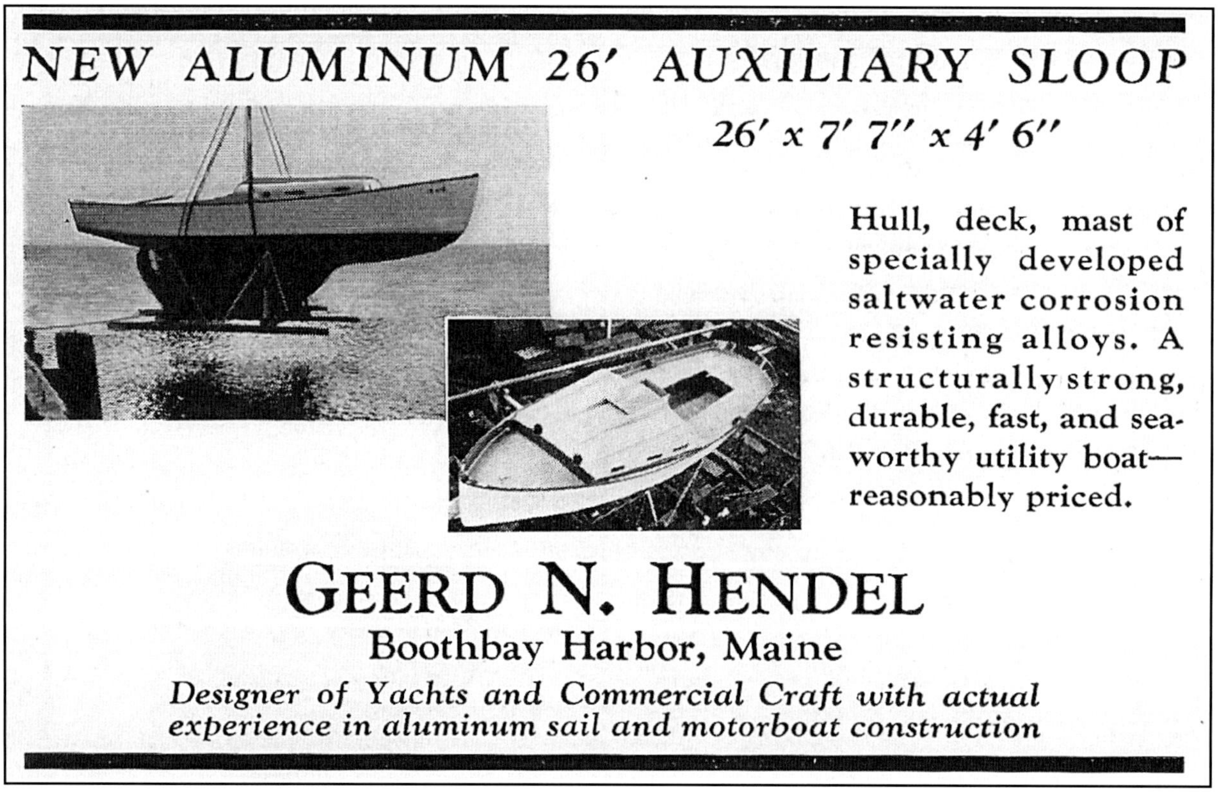

Hendel did offer Whistler *for sale almost immediately upon her completion, as evidenced by this 1940 magazine ad.*

for them to live aboard, and it was also in rough shape. Seth and his sister paid $3,000 or $3,500 for it, which was the amount of the storage fees owed. Seth became excited about the sloop's history, but he and his brother Jonathan agreed that living aboard on a year-round basis was not practical. The keel was wood with a lead shoe, and the keel (steel) bolts had corroded some of the aluminum hull flanges through electrolysis. The keel was cut off to assess the needed repairs. An attempt to weld a small patch on the hull was too difficult for the welder because of its curved shape, and that's where the repair project stopped.

Yentes hauled the sloop home to Monroe where he grew up, stored it in a hilltop field, and obtained copies of *Whistler*'s blueprints from the Geerd Hendel Collection at the Maine Maritime Museum in Bath to guide his restoration. He especially appreciated the *Whistler* because of its very aesthetically pleasing lines. His plan to restore the vessel and sail it was derailed when he purchased some farmland, met his partner, had two children, and started a business.

Currently, the hull paint is badly flaking, and the boat needs significant restoration work. The interior was stripped because it was really deteriorated. It appears that a small bowsprit was added at some point, probably to allow for a bigger headsail, but it has been removed.

A marine surveyor determined in December 2018 that "The aluminum in the aluminum hull was found to be in remarkably good condition considering her age and use in salt water, except around the floors and plate that joined to the keel."

"Whistler" was designed for his own use by Geerd M. Hendel, of Boothbay Harbor, Maine, and built by Rice Brothers, of East Boothbay. She is 26′ 2″ in length over all, 7′ 7″ beam and 4′ 6″ draft, and is built of aluminum alloy except for a wooden fin keel. A 7-18 horse power Red Wing engine furnishes auxiliary power.

Yachting magazine, August 1940.

26′ GEERD HENDEL aluminum auxiliary sloop, built by RICE BROS. yard in E. Boothbay 1939. virtually "like new" in every respect. 2 cyl. inboard, dacs, stainless rigging, two berths, head, stove and sink. extra large cockpit for day sailing, fiberglass dinghy. Offered at $4000.

Hendel listed Whistler *for sale in the Boothbay Register for $4,000 in October 1957, although she apparently remained in his ownership until 1974.*

Whistler *in April 2011, stored in a field in Monroe, Maine.*
Seth Yentes Collection.

The 1935, *Wind Call* appears to have been the next pleasure craft constructed of aluminum alloy and intended for saltwater use. When Burgess and Stephens needed a naval architect and two excellent draftsmen to prepare the plans for the *Ranger,* they first recruited G. Gilbert Wyland, who was living in New Jersey, from the New York Shipbuilding Company. Wyland had graduated about ten years before from the Webb Institute of Naval Architecture and was an avid sailor. Draftsmen Geerd Hendel and Henry Mank were then borrowed from BIW's drafting room.[74]

"The work at Bath with aluminum convinced both Gil Wyland and Geerd Hendel that aluminum could withstand corrosion and would become useful as a yacht-building material. Both built small aluminum sailing boats for their own use as soon as they could. Gil's *Wind Call,* built at the Jakobson yard in Oyster Bay, was ready for sea trials on New Year's Day, 1946....*Wind Call* was a great success, and Gil enjoyed her for more than 30 years of cruising and racing, proving that even so early a riveted aluminum hull could withstand the corrosive properties of salt water. Later in her life, some rivets were replaced, and she had extensive deck work done, but she depreciated no more than the best wood construction and less than any but the best."[75]

In a letter dated February 10, 1959, Hendel explained the addition of the bowsprit:

> In 1940, when my boat was new, friends coaxed me to enter the Monhegan Race. I had no bowsprit at that time, and the small jib and mainsail only, and knew the boat was underrigged for light winds. So, over the telephone I ordered a larger jib, practically the size and shape of the mainsail, which I planned to set flying on a housing bowsprit and also use as a spinnaker. It was not a satisfactory arrangement, but later I added a permanent bowsprit and after that used this large jib constantly, because it made a sailer out of the boat. Since it was not cut as a mainsail and was large, it rode close to the deck. But it had one fault. Sailing in a stiff breeze close hault [*sic*], frequently a wave would splash into it, and solid water would shoot up in it some 10 feet and more. It was quite a sight and lent zest to sailing. But when this jib got older, I often wondered when the wave would come which would part it. It never came, but when a new suit of sails was needed a few years ago, the jib was cut higher, which eliminated the water problem. Nevertheless, if I carried two large jibs on the boat, one of them would be cut low.[73]

In a *Boating* magazine article in 1966, G. Gilbert Wyland, then the chief designer for Sparkman & Stephens, described the condition of his then 20-year-old sloop in great detail. The vessel was still sound, but there was some minor corrosion, which he evaluated and described changes and do's and don'ts for the construction of aluminum vessels in the future.

He also described the boat's history, saying, "Just before the end of World War II, Irving Jakobson of Jakobson Shipbuilding in Oyster Bay, New York, conceived the idea of producing a stock aluminum cruising sailboat about 35 feet long. Certain aluminum alloys had been used in a limited context and had shown considerable promise for marine construction. [In 1941, the Philadelphia Navy Yard began construction of two mostly aluminum PT boats (*PT-7* and *PT-8*), designed at the Navy Yard and the Navy Bureau of Ships.] ...[A]nd yacht designer Geerd Hendel had constructed a 24' [*sic*] aluminum sloop *Whistler* for his own use."[76]

Actually, three 35' aluminum sister sloops were built by Jakobson from designs by Sparkman & Stephens: *Wind Call* for Wyland, *Dousabell* for Jakobson himself, and *Nimrod II* for the Grumman Aircraft Engineering Corporation. *Wind Call* was the prototype.

Plating for all three boats was 3/16" aluminum, and 3/8" rivets were countersunk from the outside. Decks were hard-faced plywood with wood intermediate beams where hull frames or bulkheads occurred. In contrast, *Whistler*'s decks were aluminum.

At the time when these three boats were built, as with *Whistler*, welding technology had not reached a state of reliability, so welding was used sparingly in a few areas that were not highly stressed. One of the few places where welding was used on *Wind Call* was along the trailing edge of the rudder.

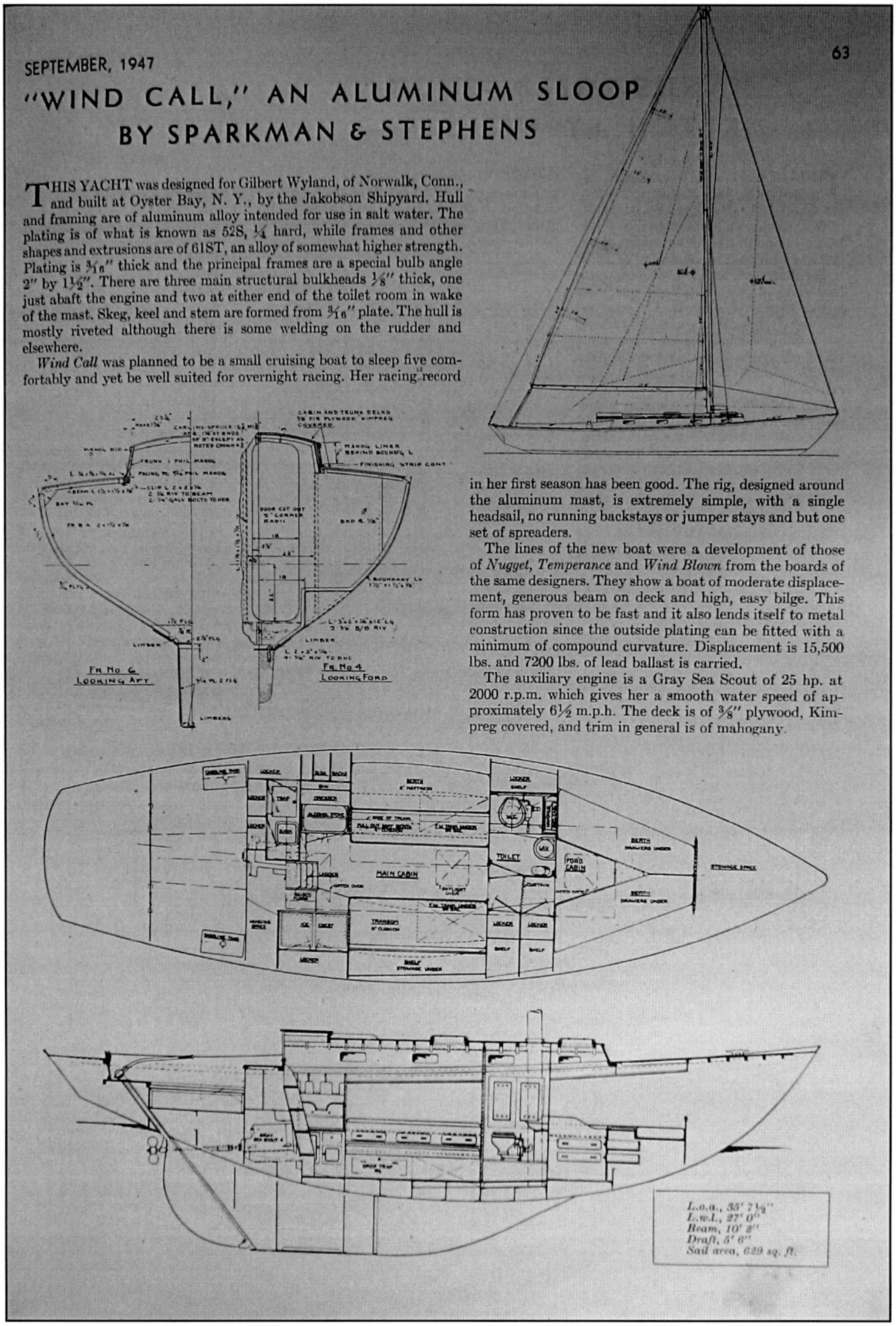

SEPTEMBER, 1947 63

"WIND CALL," AN ALUMINUM SLOOP BY SPARKMAN & STEPHENS

THIS YACHT was designed for Gilbert Wyland, of Norwalk, Conn., and built at Oyster Bay, N. Y., by the Jakobson Shipyard. Hull and framing are of aluminum alloy intended for use in salt water. The plating is of what is known as 52S, ¼ hard, while frames and other shapes and extrusions are of 61ST, an alloy of somewhat higher strength. Plating is $\frac{3}{16}$" thick and the principal frames are a special bulb angle 2" by 1½". There are three main structural bulkheads ⅛" thick, one just abaft the engine and two at either end of the toilet room in wake of the mast. Skeg, keel and stem are formed from $\frac{3}{16}$" plate. The hull is mostly riveted although there is some welding on the rudder and elsewhere.

Wind Call was planned to be a small cruising boat to sleep five comfortably and yet be well suited for overnight racing. Her racing record in her first season has been good. The rig, designed around the aluminum mast, is extremely simple, with a single headsail, no running backstays or jumper stays and but one set of spreaders.

The lines of the new boat were a development of those of *Nugget*, *Temperance* and *Wind Blown* from the boards of the same designers. They show a boat of moderate displacement, generous beam on deck and high, easy bilge. This form has proven to be fast and it also lends itself to metal construction since the outside plating can be fitted with a minimum of compound curvature. Displacement is 15,500 lbs. and 7200 lbs. of lead ballast is carried.

The auxiliary engine is a Gray Sea Scout of 25 hp. at 2000 r.p.m. which gives her a smooth water speed of approximately 6½ m.p.h. The deck is of ⅜" plywood, Kimpreg covered, and trim in general is of mahogany.

L.o.a., 35' 7½"
L.w.l., 27' 0"
Beam, 10' 2"
Draft, 5' 6"
Sail area, 629 sq. ft.

Yachting magazine, September 1947.

Wyland reported that in 20 years of use as a charter, she had never leaked except twice at the skeg (in a separate compartment) when two groundings took place.

Wyland ends his very detailed article with the following conclusion: "To make a long story short, if we had known in 1946 what we know now, and if we had the equipment, *Wind Call* would have been welded, the lead keel would be inside, and perhaps the decks would have been aluminum....But then I'm sure that 20 years hence we wouldn't do it that way, either..."[77]

Although it seems clear that *Whistler* preceded *Wind Call* by five years and hence was the first aluminum yacht built in the U.S. for salt water, there remains some occasional controversy over whether *Whistler* was actually the first. The Starling Burgess–designed 16' aluminum-alloy centerboard sloop *Pirouette* was built by ALCOA at New Kensington, Pennsylvania, in the spring of 1939 and sailed that summer. *Whistler* was designed by Hendel in early 1939, and her construction at Rice Brothers in East Boothbay began in August 1939, although she wasn't launched until the spring of 1940. The scope and scale of these two vessels are distinctly different, however. The 16' *Pirouette* weighed only 217 lbs. and carried 125 square feet of sail. She was trailerable and could be powered by an outboard motor. By contrast, *Whistler*'s total displacement was 5200 lbs., she had a fixed keel, carried a sail area of 300 square feet, and was designed for extended sailing with accommodations for two. *Whistler* was a "yacht"—using the term as it was understood in the 1930s—and is thus regarded as the first of its kind constructed from aluminum.

Note on welding aluminum: During the 1920s, major advances were made in welding technology, including the 1920 introduction of automatic welding in which electrode wire was fed continuously. Shielding gas became a subject receiving much attention, as scientists attempted to protect welds from the effects of oxygen and nitrogen in the atmosphere. Porosity and brittleness were the primary problems, and the solutions that developed included the use of hydrogen, argon, and helium as welding atmospheres. A major expansion of arc welding occurred during the 1930s and then during World War II.[78]

The Heliarc Gas Tungsten Arc Welding process (GTAW), developed in 1944, is still extensively used to successfully weld aluminum alloys today. The GTAW process can be used to weld aluminum of all thicknesses and joint types. GTAW has since been largely supplanted by the gas metal arc welding (GMAW) process for many aluminum welding applications, primarily because of the increased speed to weld thicker sections. GTAW still has an important place in the aluminum welding industry, however, and is now most often used to weld thinner gauges of aluminum (up to ¼) and for applications where aesthetics are most important.[79]

Pirouette was an especially interesting vessel from the point of view of the construction technique developed in 1937 by Burgess's brother Charles, a prominent structural engineer. Employing it, "Each side of her bottom is formed from a single sheet of metal cut as if from a huge cone, the curvature beginning with an almost flatness at the stern and increasing forward to full rounded sections at the bow." She also was constructed without ribs; transverse stiffness was supplied by the bulkheads riveted to the bottom, deck, and sides at either end of the cockpit.[80]

In his 1988 audio interview, Hendel mentioned the 26' aluminum sloop Duchess, *built in 1949 by the Aluminum Boat Company of California, as the first aluminum production boat being marketed nationally. He had clipped this ad to place in his files.*

CHAPTER SEVEN

Yachts

Hendel's interest in designing large yachts was sparked during his employment at BIW. Between 1928 and 1930, the years just before the Depression began, the yard had a remarkable number of contracts and was building mostly yachts of 205' to 250' in length, with one being completed every month. BIW's drafting team at that point consisted of twenty individuals who prepared design detail for these vessels. Hendel was hired as a draftsman by BIW on July 29, 1929.

The earliest yacht for which Hendel had a role as a BIW draftsman appears to be the 105' twin-screw motor yacht *Althea* (hull #132). The 142-ton *Althea* was built for C.H. Murphy and launched July 29, 1930, with Hendel preparing design details for bow chocks, bulwark doublers, and strakes. She was designed by Henry J. Gielow, Inc., who were naval architects and yacht brokers. The company's main office was at 25 West 43rd Street in New York City, with other offices in Los Angeles, Miami, Detroit, and Boston.

1930 Althea.

Geerd Hendel archives, Herbert Douglas photo, Maine Maritime Museum.

Hendel noted in his 1988 oral interview that he had assisted with developing the detailed design plans for the Gielow-designed Hi-Esmaro, *a 270' yacht of 1200 tons with a cruising range of 15,000 miles. She carried a crew of 45.*

Geerd Hendel archives, Herbert Douglas photo, Maine Maritime Museum.

1931 Halonia, *a 147' diesel-powered Gielow-designed yacht built for Charles A. Thorne, was a third yacht for which Hendel prepared design details at BIW.*

Geerd Hendel archives, Herbert Douglas photo, Maine Maritime Museum.

The late 1920s were a very heady time for yacht designers, as buyers did not anticipate the coming economic depression, the initial phase of which occurred in 1929–31.

1952 *Sooner*: Hendel designed a 68' diesel trawler-yacht constructed at Camden Shipbuilding for Julian Davis (see Chapter 10).

1952 *Porpoise*, a 61' sport fisherman, which Paterson, New Jersey, industrialist Allison Laytham had commissioned Hendel to design, was built by Camden Shipbuilding Company, Inc. *Porpoise* was featured on the cover of *The Rudder* magazine and is regarded by several design experts as one of the handsomest vessels ever conceived (see Chapter 10).

Hendel's 1932 design for a 55' steel yacht lacked the elegance of his future power yacht designs, and there's no record that a yacht of this design was ever built.

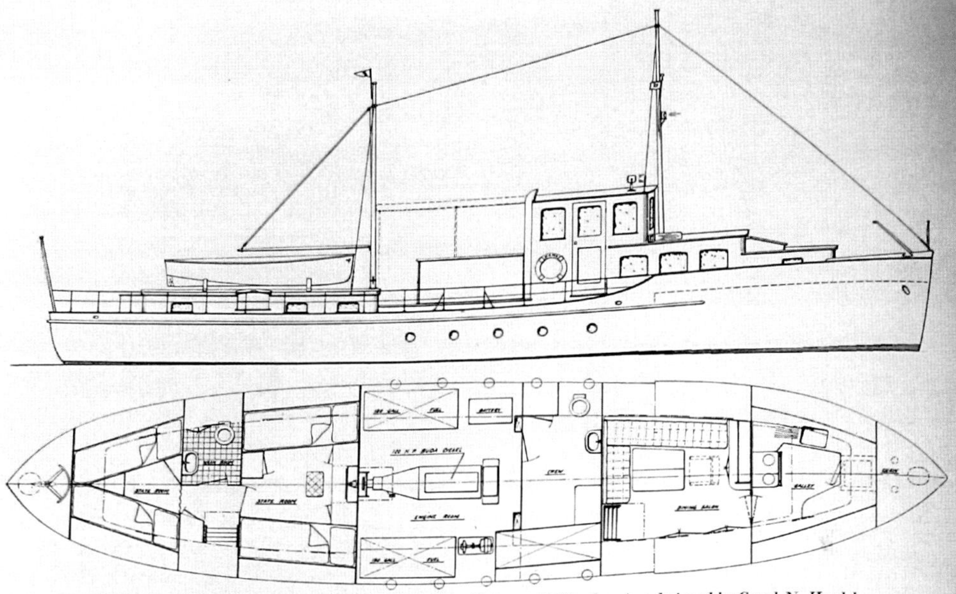

Outboard profile and interior accommodation plan of the 55-foot steel Diesel cruiser designed by Geerd N. Hendel

Comfort and Seaworthiness in This 55-Foot Steel Yacht

ROOMINESS, comfort, and sea-keeping ability are the keynotes of the design of the 55-foot steel Diesel cruiser whose plans are shown above. She is from the board of Geerd N. Hendel, of Boothbay Harbor, Maine, who has gone to great pains to produce a boat that was not only seaworthy, but that would have as easy a motion as possible in bad going, and an easy form to drive. Her beam is 12 feet and draft 4 feet, and a 4-cylinder Buda Diesel with 2:1 reduction gear is expected to drive her 12 miles an hour in average going.

The excellent interior accommodations are clearly shown in the plans. Great care has been taken in the design to prevent any sharp bends in frames, and with the advent of welding steel plates in place of riveting and caulking, construction cost rivals that of a wooden boat.

The Rudder magazine, August 1932.

In 1953–55, Hendel designed *Vedersein* (later *Salamar, Amerada, Cygnus, Genie,* and *Sans Souci*). *Vedersein* is an American adaption of the German *Wiedersehen,* meaning *meet again.* The 110' yacht built by Abeking & Rasmussen in Germany for Allison Laytham, had a maximum beam of 19' and a draft of 6.5'. Engine output was 1520 hp, its cruising speed was 12 knots, and its nautical range was 1500 miles.

"*Vedersein* has been lauded as one of the most handsome and well-proportioned yachts ever built, and her mahogany hull may make her the largest wood-hulled motor yacht ever constructed."[81]

Current records show her wooden hull size record was surpassed by that of the 127.12' *Alyssa* (later *Fundamental*) built in Turkey in 2014.

Marine photographer Peter Barlow described Hendel's super-yacht designs as having a relatively small superstructure for their long, lean hulls, plus modest streamlining and a timeless quality, all common themes of the designer's work.

Vedersein.
Geerd Hendel Collection, courtesy of his stepdaughter, Brigitte M. Gefrich.

Motor Yacht Vedersein *refitted in 1986 and renamed Salamar.*
Geerd Hendel Collection, courtesy of his stepdaughter, Brigitte M. Gefrich.

Vedersein *being loaded at the Abeking & Rasmussen yard in Germany for shipment to the United States.*
Geerd Hendel Collection, courtesy of his stepdaughter, Brigitte M. Gefrich.

In 1963, *Vedersein* was renamed *Salamar*, received upgrading, and *MotorBoating* magazine's September 1964 issue described the changes as follows: "When she was brought to Port Everglades last December, she was rechristened *Salamar* and treated to a refurbishing in keeping with her standards of quality, including new electronic gear, a new power plant and a change from her heavily Teutonic furnishings to functional and pleasing American-style wares."

As *Sans Souci* in 1986, the former *Vedersein* was re-powered with two Caterpillar D353 425-hp diesel engines, making her capable of a top speed of 14 knots and cruising comfortably at 12 knots. With her 3300-gallon fuel tanks, she had a maximum range of 1500 nautical miles. Her water tanks stored around 1200-gallons of fresh water.

1956 yacht *Admiral Blake*, a 73' yacht built by Camden Shipbuilding Company, Inc. (hull Y-16; see Chapter 10).

1970–71 luxury yacht *Claybeth* (later *George*, *Challenger*, *Sharon S.*); 124' for Allison Laytham.

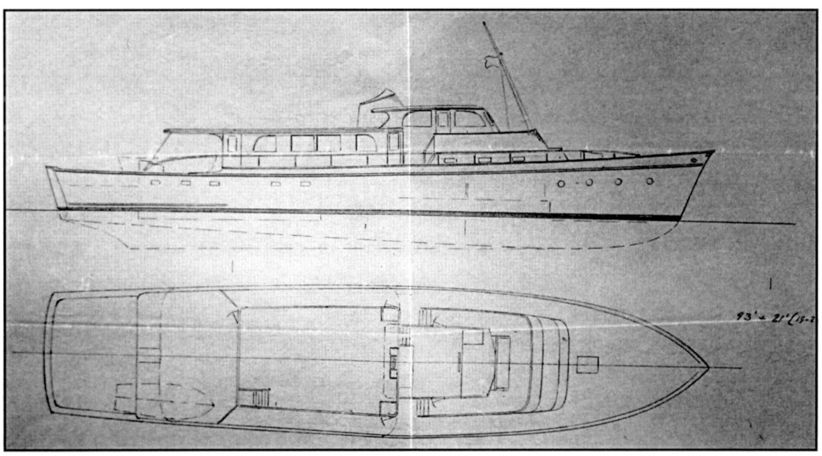

Hendel's undated sketch plan for a 93' yacht with a 21' beam, similar in design to Vedersein.
Geerd Hendel Collection, MS-310, Maine Maritime Museum.

Claybeth, *designed by Geerd Hendel.*
Hendel Photograph Collection, MS-310, Box 1, Folder 16, Maine Maritime Museum.

Claybeth (as *George*) is a 124' (37.8 m) motor yacht with a beam of 22'11" (7.0 m) and a 6'9" (2.07 m) draft. Custom built in 1970 by Abeking & Rasmussen; the luxury vessel's sophisticated exterior design and engineering are the work of Geerd Hendel. As *Claybeth,* she was built for E. Clayton Gengras of Hartford, Connecticut, chairman and president of the Security Insurance Corporation.

As *George,* last refitted in 2005, she was described as having a steel hull, wood superstructure, a sandalwood interior, and three teak decks. The custom semi-displacement yacht is equipped with an ultra-modern stabilization system that reduces roll motion effect and results in a smoother, more enjoyable cruising experience.

With two Caterpillar 760 diesel engines, *George* is capable of 12 knots flat-out, with a cruising speed of 10 knots and fuel capacity of 6800 gallons.

The vessel offers accommodation for up to ten guests in six suites comprising one owner's cabin, three double cabins, and two twin cabins. She is also capable of carrying up to seven crew members.[82]

Claybeth *as* George *specifications:*

Overview	
Name:	George
Type:	Motor Yacht
Model:	Custom

Builder:	Abeking & Rasmussen
Naval Architect:	Geerd N. Hendel
Exterior Designers:	Geerd N. Hendel
Interior Designer:	
Year:	1970 / 2005
Hull NB:	6338
Hull Colour:	-

Dimensions	
Length Overall:	37.80m (124' 0" ft)
Length at Waterline:	34.50m (113' 2" ft)
Beam:	7.00m (22' 11" ft)
Draft (min):	-
Draft (max):	2.07m (6' 9" ft)
Gross Tonnage:	4 tons

Accommodations	
Guests:	10
Cabins Total:	6
Cabins:	1 Master, 3 Double,
	and 2 twin
Crew:	7

Construction	
Hull Configuration:	Semi-displacement
Hull Material:	Steel
Superstructure:	GRP / Wood
Deck Material:	Teak
Decks NB:	3

Engine(s)	
Quantity:	2
Fuel Type:	Diesel
Manufacturer:	Caterpillar, Inc.
Model:	D348
Power:	760 h.p. / 559kW
Total Power:	1520 h.p. / 1118kW
Propulsion:	Twin Screw

Performance & Capabilities	
Max Speed:	12.00 kts
Cruising Speed:	10.00 kts
Range:	-
Fuel Capacity:	31,037.00 L
	(6,827.18 USG)
Water Capacity:	5,300.00 L
	(1,165.84 USG)

Equipment	
Generator:	2 x Onan 65 kW
Stabilizers:	Vosper

Claybeth *(designed by DeVooght Naval Architects).*
From the website of Feadship, Royal Dutch Shipyards.

It should be noted that E. Clayton Gengras, six years after the construction of the Hendel-designed *Claybeth,* had a second *Claybeth* built in 1978 by the Feadship Royal Dutch Shipyards in the Netherlands (later *The Highlander IV, Virginian, Lady Allison, Sirenuse,* and *Berilda*). Designed by De Vooght Naval Architects, she was 126' LOA. Information about these two *Claybeths* is often confused or blended.

The 1974 *Aweigh* (later *Brasil*) is a 61' motor yacht built for Fred Requardt. Original construction is 5/4" strip plank mahogany sheathed in an epoxy/glass barrier and an impervious copper bottom. It is now powered by Cummins N-855-M diesels with 800-gallon fuel tanks and has a 12-kW Witte diesel generator. It features Arcturus Bow Thrusters, hydraulic winches, and a windlass with two 100-lb. Danforth anchors. The yacht is designed for short-handed cruising and as a live-aboard.

Aweigh.
Hendel Photograph Collection, Box 1, Folder 8, Maine Maritime Museum.

As Brasil *in 2019: 45 years after her launching, the vessel's exterior is virtually unchanged, and she is now operated for charter in Providence, Rhode Island.*

CHAPTER EIGHT

Cruising Sloops, Cutters, Ketches, and Yawls

Hendel designed a 33' auxiliary cruising ketch in 1932, using this half-hull model that he crafted himself. There are no records that describe this vessel as being built.

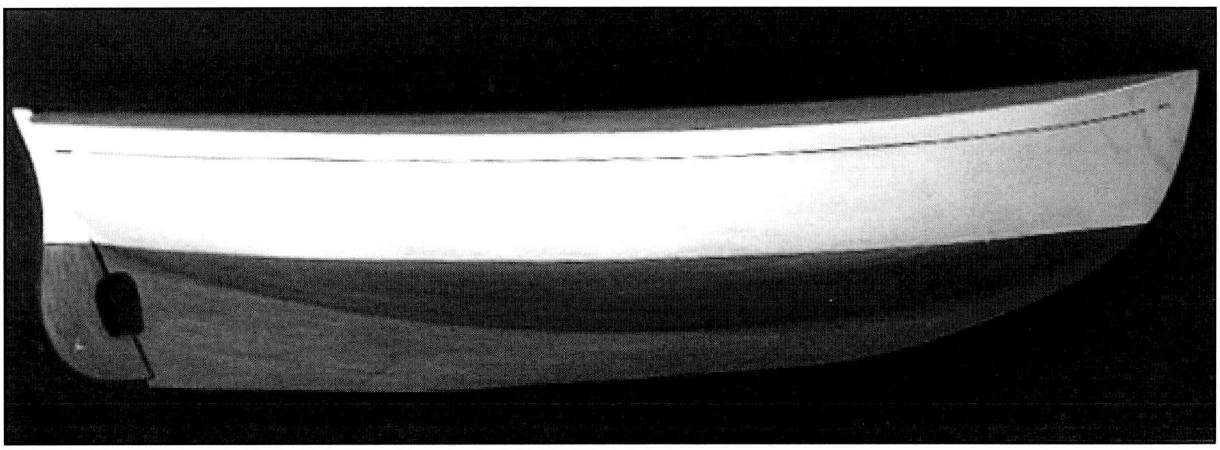

33' ketch designed by Geerd Hendel. Scale: ½" = 1'0"; model by Geerd Hendel.
Bartoldt and Ingrid Brummer Collection.

In 1935, Hendel designed a 26' sloop for Mr. Braine, a man whom he had taught to sail.

By 1936, Hendel's naval architecture practice had become active. The *Boothbay Register* edition of February 21, 1936, contained a brief article noting that Geerd Hendel, naval architect, had been drafting plans for a 33' boat at Hodgdon Bros. yard. Goudy & Stevens/Hodgdon Bros. hull records, however, aren't clear whether this boat was built.

The Boothbay Register's August 1936 "Motor Boating News" report describes three boats of his design under construction.

> AUGUST, 1936
>
> **Geerd N. Hendel**
>
> SEVERAL new craft are building from plans by Geerd N. Hendel, Boothbay Harbor, Me. There is a 24-foot auxiliary sloop, owned by Chas. H. Sprague, Darien, Conn. Builder is Norman Hodgdon, Jr., Boothbay Harbor, Maine. Dimensions are 24 feet by 21 feet 6 inches by 7 feet 9 inches by 3 feet 6 inches. Outside ballast 1750 pounds cast iron, no inside ballast. Marconi rig with 345 square feet of sail, no bowsprit. Motor: 5/10 h.p. Universal Blue Jacket Twin. A short ended, Philippine mahogany planked boat, with curved transom stern and outboard rudder, trunk cabin with 4 feet 8 inches headroom, accommodations for two, galley facilities, ice box, pump toilet, 7-foot self-bailing cockpit. The craft was designed for day sailing and short cruises.
>
> A 25-foot shoal draft auxiliary sloop is being built for Major Philip S. Wood, Fort Benning, Ga., by Roger I. Sawyer, Milbridge, Maine. Dimensions are 25 feet by 22 feet 6 inches by 8 feet by 2 feet 1 inch (4 feet 6 inches with board down). Outside ballast 1750 pounds cast iron, no inside ballast. Marconi rig with 327 square feet of sail, no bowsprit. Motor: 10/16 h.p. Gray Seascout. A short ended centerboard boat, white pine planking, curved transom stern and outboard rudder, trunk cabin with 4 feet 7 inches headroom, accommodations for two, galley facilities, ice box, pump toilet, deep cockpit. This craft also was designed for day sailing and short cruises.
>
> For his own account Mr. Hendel is building a 20-foot keel sloop. Builder is Norman Hodgdon, Jr., Boothbay Harbor, Maine. Dimensions: 20 feet 4 inches by 19 feet 10 inches by 5 feet 8 inches, 3 feet 3 inches. Outside ballast 600 pounds lead. Marconi rig with 190 square feet of sail, loose footed main sail. A Philippine mahogany planked boat with long waterline, slightly curved transom stern, large, open cockpit. This sloop fits into the Boothbay 20-footer class and is designed for racing and day sailing.

- *24' auxiliary sloop* Nema.
- *A 25' shoal draft auxiliary sloop that was to be sailed on Chesapeake Bay (see plan on following page).*
- *Hendel's own Boothbay 20' class keel sloop, probably* Sanderling, *which was described in Chapter 5.*

In 1934, Hendel designed a 25' shoal draft auxiliary cruising sloop for Major Philip Wood to sail on Chesapeake Bay; she was built in 1936.

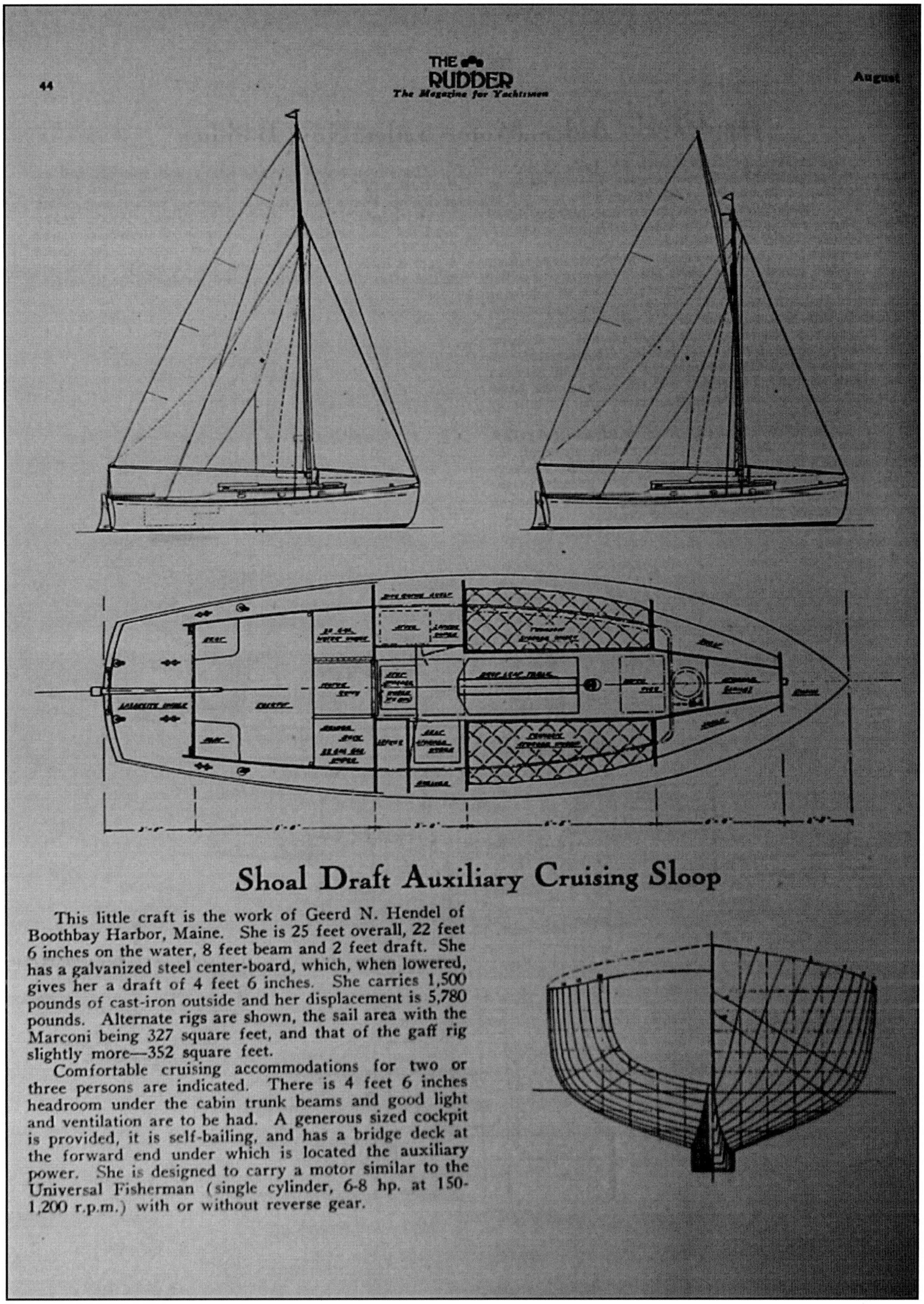

44 THE RUDDER The Magazine for Yachtsmen August

Shoal Draft Auxiliary Cruising Sloop

This little craft is the work of Geerd N. Hendel of Boothbay Harbor, Maine. She is 25 feet overall, 22 feet 6 inches on the water, 8 feet beam and 2 feet draft. She has a galvanized steel center-board, which, when lowered, gives her a draft of 4 feet 6 inches. She carries 1,500 pounds of cast-iron outside and her displacement is 5,780 pounds. Alternate rigs are shown, the sail area with the Marconi being 327 square feet, and that of the gaff rig slightly more—352 square feet.

Comfortable cruising accommodations for two or three persons are indicated. There is 4 feet 6 inches headroom under the cabin trunk beams and good light and ventilation are to be had. A generous sized cockpit is provided, it is self-bailing, and has a bridge deck at the forward end under which is located the auxiliary power. She is designed to carry a motor similar to the Universal Fisherman (single cylinder, 6-8 hp. at 150-1,200 r.p.m.) with or without reverse gear.

The Rudder Magazine, March 1934.

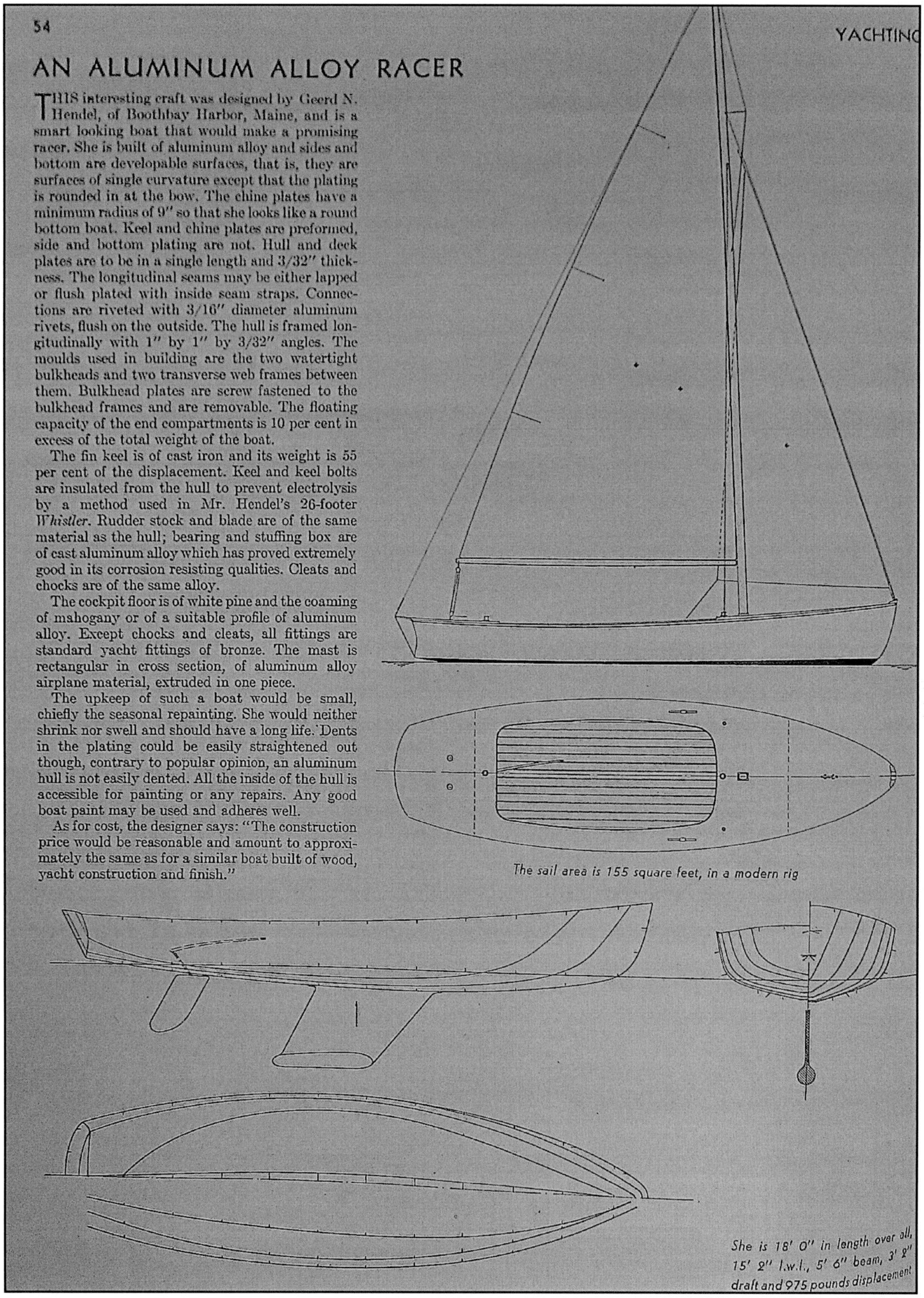

54 YACHTING

AN ALUMINUM ALLOY RACER

THIS interesting craft was designed by Geerd N. Hendel, of Boothbay Harbor, Maine, and is a smart looking boat that would make a promising racer. She is built of aluminum alloy and sides and bottom are developable surfaces, that is, they are surfaces of single curvature except that the plating is rounded in at the bow. The chine plates have a minimum radius of 9″ so that she looks like a round bottom boat. Keel and chine plates are preformed, side and bottom plating are not. Hull and deck plates are to be in a single length and 3/32″ thickness. The longitudinal seams may be either lapped or flush plated with inside seam straps. Connections are riveted with 3/16″ diameter aluminum rivets, flush on the outside. The hull is framed longitudinally with 1″ by 1″ by 3/32″ angles. The moulds used in building are the two watertight bulkheads and two transverse web frames between them. Bulkhead plates are screw fastened to the bulkhead frames and are removable. The floating capacity of the end compartments is 10 per cent in excess of the total weight of the boat.

The fin keel is of cast iron and its weight is 55 per cent of the displacement. Keel and keel bolts are insulated from the hull to prevent electrolysis by a method used in Mr. Hendel's 26-footer *Whistler*. Rudder stock and blade are of the same material as the hull; bearing and stuffing box are of cast aluminum alloy which has proved extremely good in its corrosion resisting qualities. Cleats and chocks are of the same alloy.

The cockpit floor is of white pine and the coaming of mahogany or of a suitable profile of aluminum alloy. Except chocks and cleats, all fittings are standard yacht fittings of bronze. The mast is rectangular in cross section, of aluminum alloy airplane material, extruded in one piece.

The upkeep of such a boat would be small, chiefly the seasonal repainting. She would neither shrink nor swell and should have a long life. Dents in the plating could be easily straightened out though, contrary to popular opinion, an aluminum hull is not easily dented. All the inside of the hull is accessible for painting or any repairs. Any good boat paint may be used and adheres well.

As for cost, the designer says: "The construction price would be reasonable and amount to approximately the same as for a similar boat built of wood, yacht construction and finish."

The sail area is 155 square feet, in a modern rig

She is 18′ 0″ in length over all, 15′ 2″ l.w.l., 5′ 6″ beam, 3′ 2″ draft and 975 pounds displacement

Hendel's 1940 design for an 18′ aluminum racing sloop, some of which were built at Camden Shipbuilding.

Yachting magazine, August 1940.

1940: An 18' aluminum racing sloop under sail in Camden Harbor.
Barbara Dyer Collection, Walsh History Center, Camden Public Library.

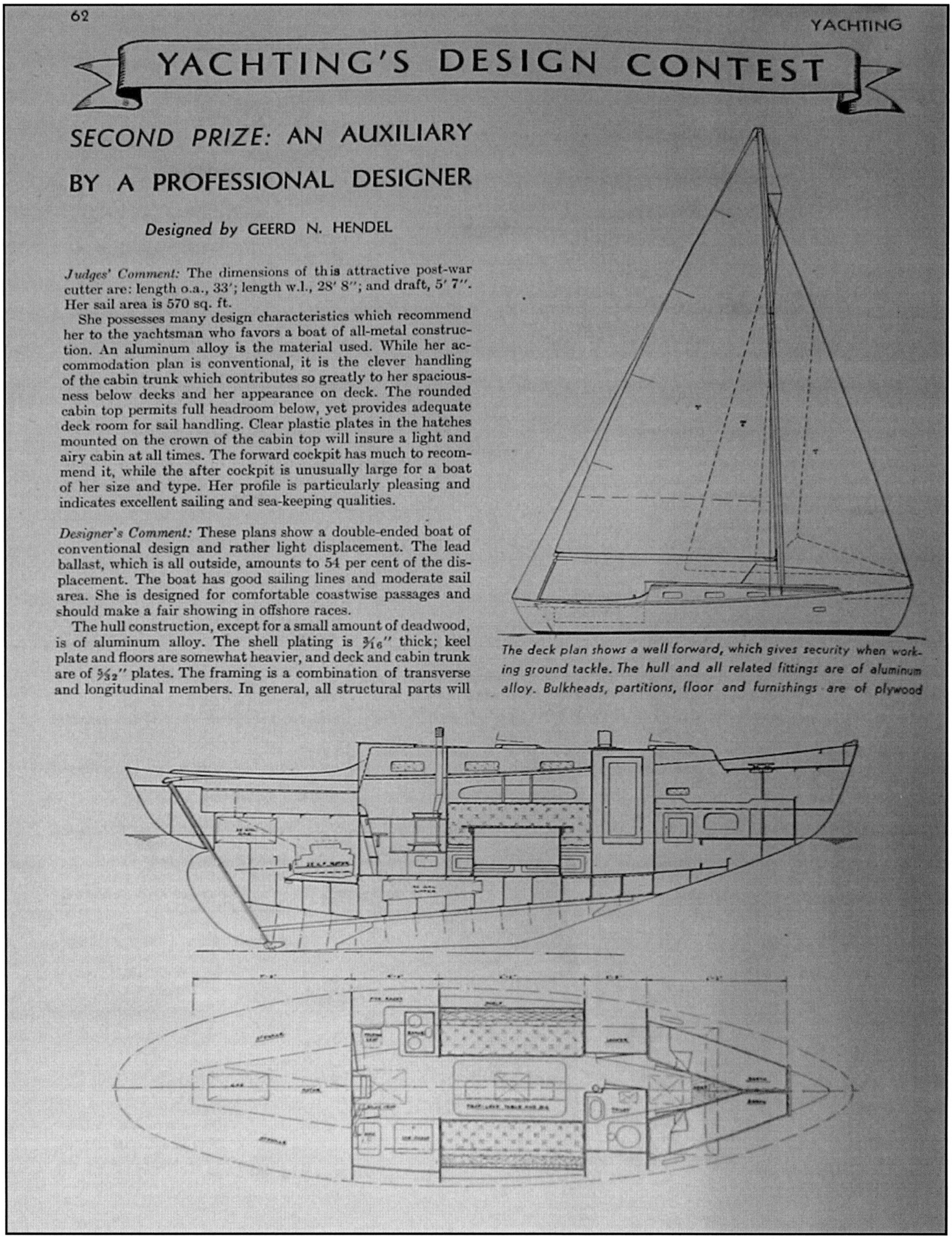

YACHTING'S DESIGN CONTEST

SECOND PRIZE: AN AUXILIARY BY A PROFESSIONAL DESIGNER

Designed by GEERD N. HENDEL

Judges' Comment: The dimensions of this attractive post-war cutter are: length o.a., 33'; length w.l., 28' 8"; and draft, 5' 7". Her sail area is 570 sq. ft.

She possesses many design characteristics which recommend her to the yachtsman who favors a boat of all-metal construction. An aluminum alloy is the material used. While her accommodation plan is conventional, it is the clever handling of the cabin trunk which contributes so greatly to her spaciousness below decks and her appearance on deck. The rounded cabin top permits full headroom below, yet provides adequate deck room for sail handling. Clear plastic plates in the hatches mounted on the crown of the cabin top will insure a light and airy cabin at all times. The forward cockpit has much to recommend it, while the after cockpit is unusually large for a boat of her size and type. Her profile is particularly pleasing and indicates excellent sailing and sea-keeping qualities.

Designer's Comment: These plans show a double-ended boat of conventional design and rather light displacement. The lead ballast, which is all outside, amounts to 54 per cent of the displacement. The boat has good sailing lines and moderate sail area. She is designed for comfortable coastwise passages and should make a fair showing in offshore races.

The hull construction, except for a small amount of deadwood, is of aluminum alloy. The shell plating is 3/16" thick; keel plate and floors are somewhat heavier, and deck and cabin trunk are of 5/32" plates. The framing is a combination of transverse and longitudinal members. In general, all structural parts will

The deck plan shows a well forward, which gives security when working ground tackle. The hull and all related fittings are of aluminum alloy. Bulkheads, partitions, floor and furnishings are of plywood

1944 aluminum alloy 33' cutter, second place winner in Yachting *magazine's August 1944 design contest.*
Yachting magazine, August 1944.

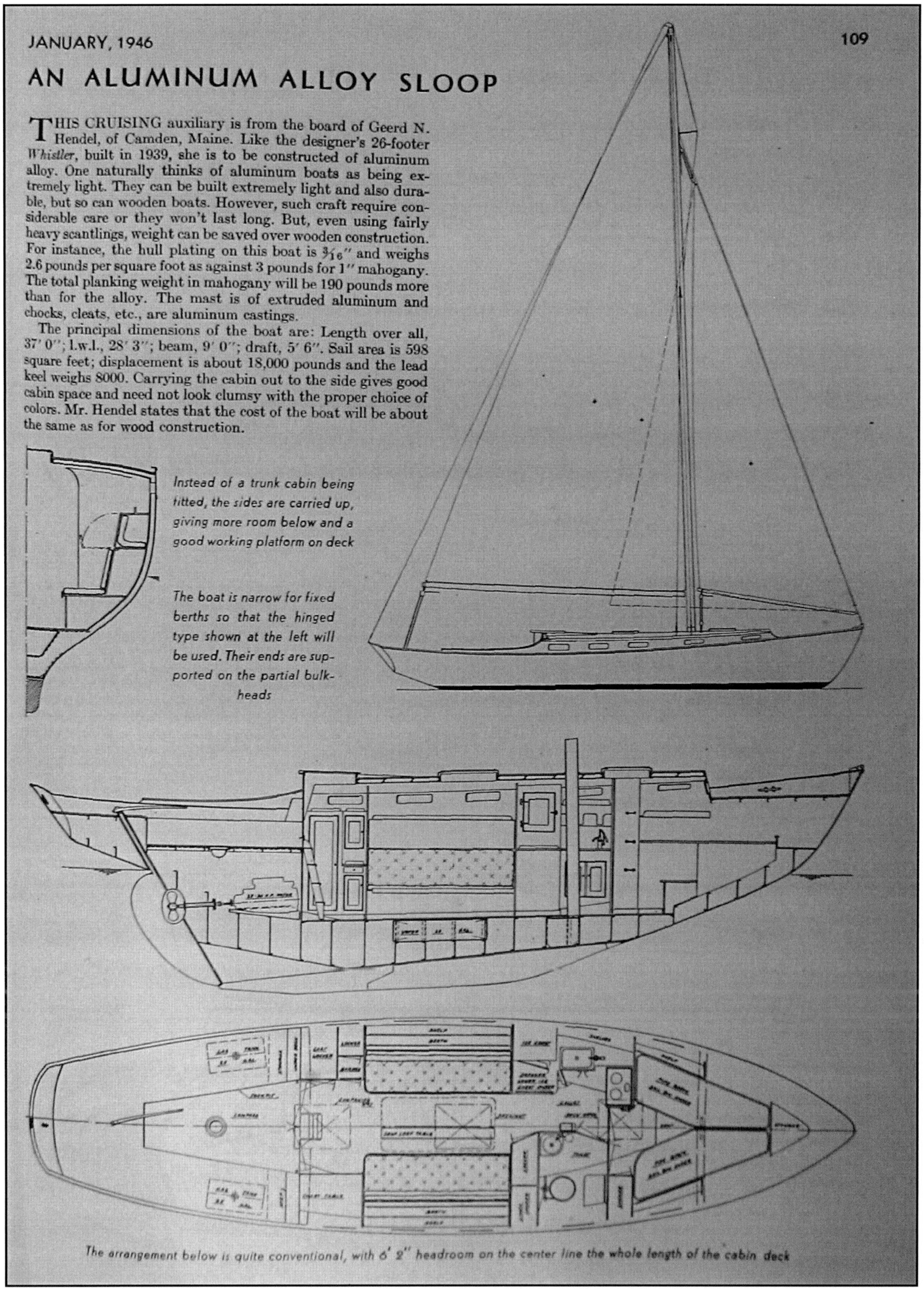

JANUARY, 1946 109

AN ALUMINUM ALLOY SLOOP

THIS CRUISING auxiliary is from the board of Geerd N. Hendel, of Camden, Maine. Like the designer's 26-footer *Whistler*, built in 1939, she is to be constructed of aluminum alloy. One naturally thinks of aluminum boats as being extremely light. They can be built extremely light and also durable, but so can wooden boats. However, such craft require considerable care or they won't last long. But, even using fairly heavy scantlings, weight can be saved over wooden construction. For instance, the hull plating on this boat is 3/16″ and weighs 2.6 pounds per square foot as against 3 pounds for 1″ mahogany. The total planking weight in mahogany will be 190 pounds more than for the alloy. The mast is of extruded aluminum and chocks, cleats, etc., are aluminum castings.

The principal dimensions of the boat are: Length over all, 37′ 0″; l.w.l., 28′ 3″; beam, 9′ 0″; draft, 5′ 6″. Sail area is 598 square feet; displacement is about 18,000 pounds and the lead keel weighs 8000. Carrying the cabin out to the side gives good cabin space and need not look clumsy with the proper choice of colors. Mr. Hendel states that the cost of the boat will be about the same as for wood construction.

Instead of a trunk cabin being fitted, the sides are carried up, giving more room below and a good working platform on deck

The boat is narrow for fixed berths so that the hinged type shown at the left will be used. Their ends are supported on the partial bulkheads

The arrangement below is quite conventional, with 6′ 2″ headroom on the center line the whole length of the cabin deck

1946 design for an aluminum 37′ auxiliary sloop.

Yachting magazine, January 1946.

1946: A 33' sloop design prepared for William "Pete" Peterson, owner of Camden Shipbuilding (see Chapter 10).

1947: *Gone Away* (later *Fair Thee Well*), a 42' sloop with a beam of 10'4" built for Morris Cheston of Philadelphia by Camden Shipbuilding Company (hull Y-6).

Her keel, backbone and frames were of white oak, and her hull was African mahogany. She was powered by a 46-hp Gray 162 Luger engine and had Merriman winches and roller reefing. She carried 845 square feet of sail, and her rigging was stainless steel.[83]

Her 63' hollow aluminum mast helped to reduce her weight and enabled her to carry full sail comfortably in a stiff breeze.

The caption of an unattributed newspaper clipping at the Maine Maritime Museum's Geerd Hendel Collection describes *Gone Away* as "having a clean profile which shows her moderate ends, fine sheer, and flaring bow. Her modern rig makes for easy handling. African mahogany planking, clear native pine decks, and stainless-steel rigging are indicative of material put into this boat."

Gone Away.

Geerd Hendel Collection, courtesy of his stepdaughter, Brigitte M. Gefrich.

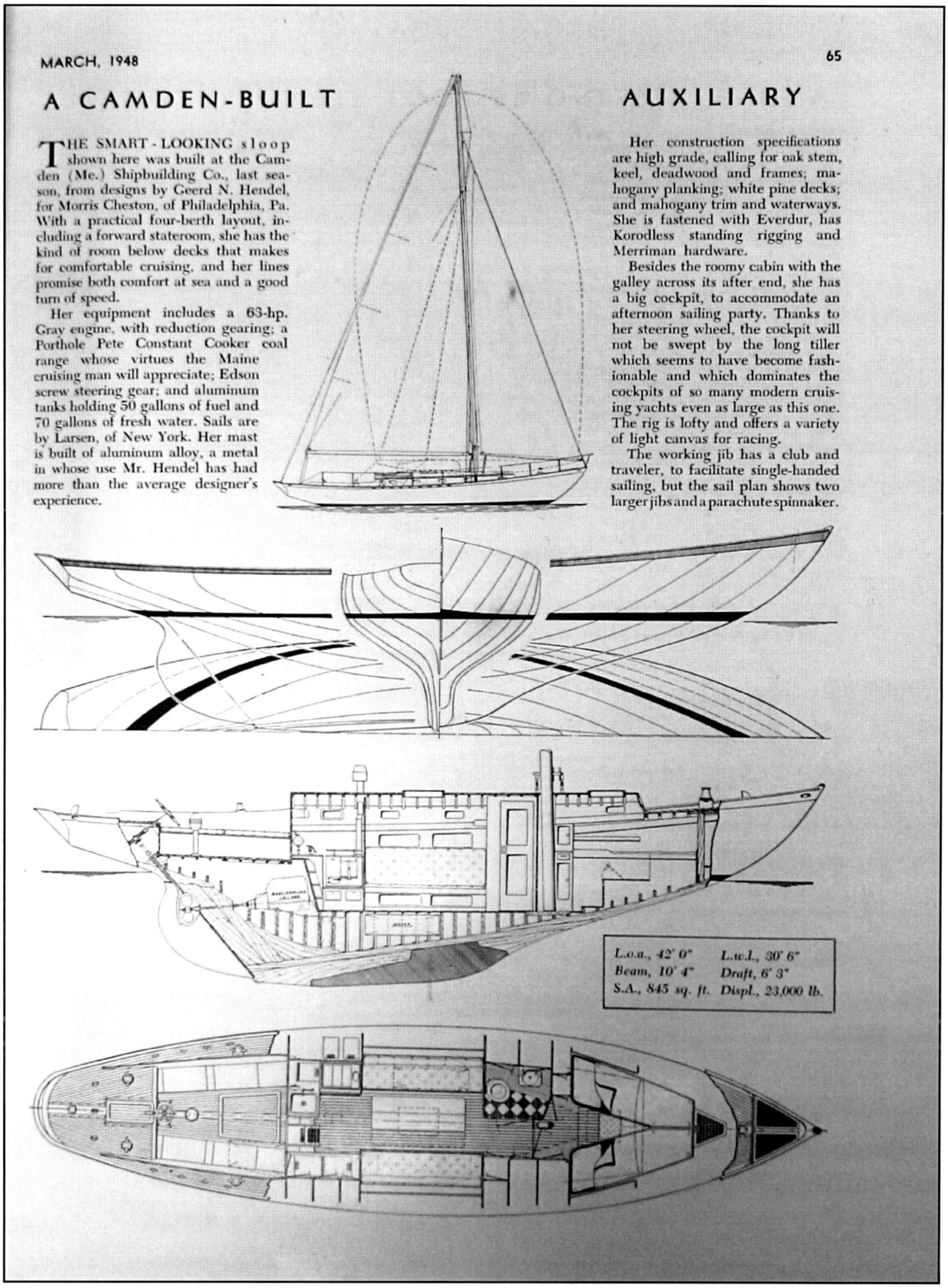

MARCH, 1948 65

A CAMDEN-BUILT AUXILIARY

THE SMART-LOOKING sloop shown here was built at the Camden (Me.) Shipbuilding Co., last season, from designs by Geerd N. Hendel, for Morris Cheston, of Philadelphia, Pa. With a practical four-berth layout, including a forward stateroom, she has the kind of room below decks that makes for comfortable cruising, and her lines promise both comfort at sea and a good turn of speed.

Her equipment includes a 63-hp. Gray engine, with reduction gearing; a Porthole Pete Constant Cooker coal range whose virtues the Maine cruising man will appreciate; Edson screw steering gear; and aluminum tanks holding 50 gallons of fuel and 70 gallons of fresh water. Sails are by Larsen, of New York. Her mast is built of aluminum alloy, a metal in whose use Mr. Hendel has had more than the average designer's experience.

Her construction specifications are high grade, calling for oak stem, keel, deadwood and frames; mahogany planking; white pine decks; and mahogany trim and waterways. She is fastened with Everdur, has Korodless standing rigging and Merriman hardware.

Besides the roomy cabin with the galley across its after end, she has a big cockpit, to accommodate an afternoon sailing party. Thanks to her steering wheel, the cockpit will not be swept by the long tiller which seems to have become fashionable and which dominates the cockpits of so many modern cruising yachts even as large as this one. The rig is lofty and offers a variety of light canvas for racing.

The working jib has a club and traveler, to facilitate single-handed sailing, but the sail plan shows two larger jibs and a parachute spinnaker.

L.o.a., 42' 0"	*L.w.l., 30' 6"*
Beam, 10' 4"	*Draft, 6' 3"*
S.A., 845 sq. ft.	*Displ., 23,000 lb.*

Gone Away's *design detail.*

Yachting magazine, March 1948.

1948: *Alphard* was a 42' sloop designed by Hendel and built by Camden Shipbuilding for Judge Curtis Bok of Philadelphia and Rockport, Maine. About 1954, she was converted to a ketch with her aluminum alloy mainmast shortened by about 8' and a mizzenmast added to make her easier to handle in a planned trans-Atlantic crossing with anticipated rough weather and a small crew (see Chapter 10).[84]

1948 Alphard *shown under sail in a Camden Shipbuilding Company ad.*

Yachting magazine, December 1948.

Godofin *under sail.*

Geerd Hendel Collection, courtesy of his stepdaughter, Brigitte M. Gefrich.

During his 1988 oral interview, Hendel noted that the hull design of 1948 *Alphard* was a modification of the 1947 *Gone Away*, and said their interior layouts were identical. In response to a question about where that particular hull shape came from, he responded that the Boothbay 20 provided his starting point. His three interviewers agreed that they regarded both the *Alphard* and *Gone Away* as visually beautiful vessels.

1948: *Godofin* was a 32' sloop, owned by Albert H. Chatfield Jr. of Rockport, Maine, and Cincinnati, Ohio, which he and his wife, Marion, sailed for pleasure in Maine waters. Albert Chatfield was the commodore of the Camden Yacht Club from 1961–63.

Godofin was described in the Design Section of the *Yachting* magazine edition of December 1948, p. 70, in a review entitled "A Cruising Sloop by Geerd Hendel." The vessel was built by the Paul E. Luke boatyard, East Boothbay, Maine.

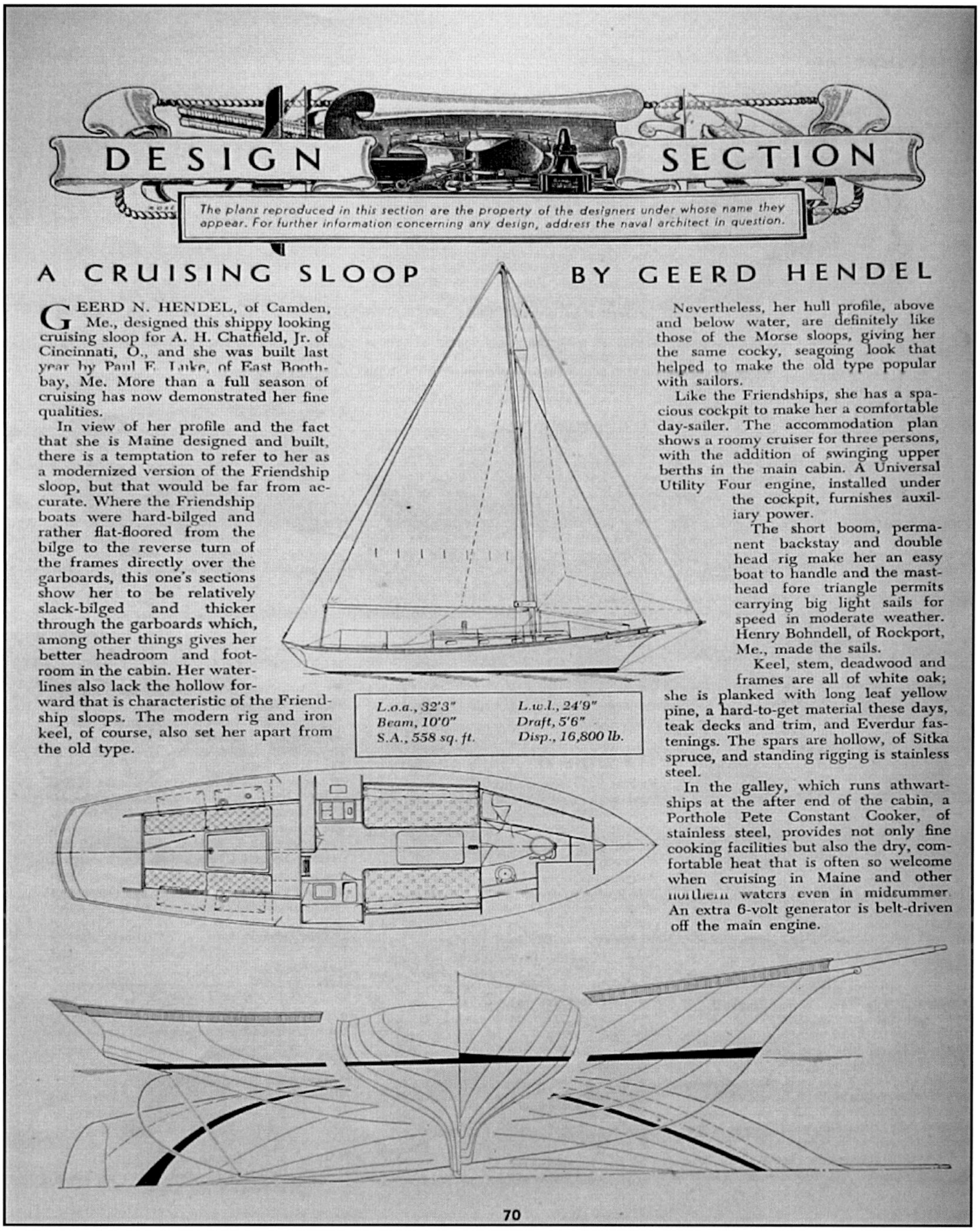

DESIGN SECTION

The plans reproduced in this section are the property of the designers under whose name they appear. For further information concerning any design, address the naval architect in question.

A CRUISING SLOOP BY GEERD HENDEL

Geerd N. Hendel, of Camden, Me., designed this shippy looking cruising sloop for A. H. Chatfield, Jr. of Cincinnati, O., and she was built last year by Paul E. Luke, of East Boothbay, Me. More than a full season of cruising has now demonstrated her fine qualities.

In view of her profile and the fact that she is Maine designed and built, there is a temptation to refer to her as a modernized version of the Friendship sloop, but that would be far from accurate. Where the Friendship boats were hard-bilged and rather flat-floored from the bilge to the reverse turn of the frames directly over the garboards, this one's sections show her to be relatively slack-bilged and thicker through the garboards which, among other things gives her better headroom and foot-room in the cabin. Her water-lines also lack the hollow forward that is characteristic of the Friendship sloops. The modern rig and iron keel, of course, also set her apart from the old type.

Nevertheless, her hull profile, above and below water, are definitely like those of the Morse sloops, giving her the same cocky, seagoing look that helped to make the old type popular with sailors.

Like the Friendships, she has a spacious cockpit to make her a comfortable day-sailer. The accommodation plan shows a roomy cruiser for three persons, with the addition of swinging upper berths in the main cabin. A Universal Utility Four engine, installed under the cockpit, furnishes auxiliary power.

The short boom, permanent backstay and double head rig make her an easy boat to handle and the mast-head fore triangle permits carrying big light sails for speed in moderate weather. Henry Bohndell, of Rockport, Me., made the sails.

Keel, stem, deadwood and frames are all of white oak; she is planked with long leaf yellow pine, a hard-to-get material these days, teak decks and trim, and Everdur fastenings. The spars are hollow, of Sitka spruce, and standing rigging is stainless steel.

In the galley, which runs athwartships at the after end of the cabin, a Porthole Pete Constant Cooker, of stainless steel, provides not only fine cooking facilities but also the dry, comfortable heat that is often so welcome when cruising in Maine and other northern waters even in midsummer. An extra 6-volt generator is belt-driven off the main engine.

L.o.a., 32'3"	*L.w.l., 24'9"*
Beam, 10'0"	*Draft, 5'6"*
S.A., 558 sq. ft.	*Disp., 16,800 lb.*

70

Godofin.

Yachting Magazine, December 1948.

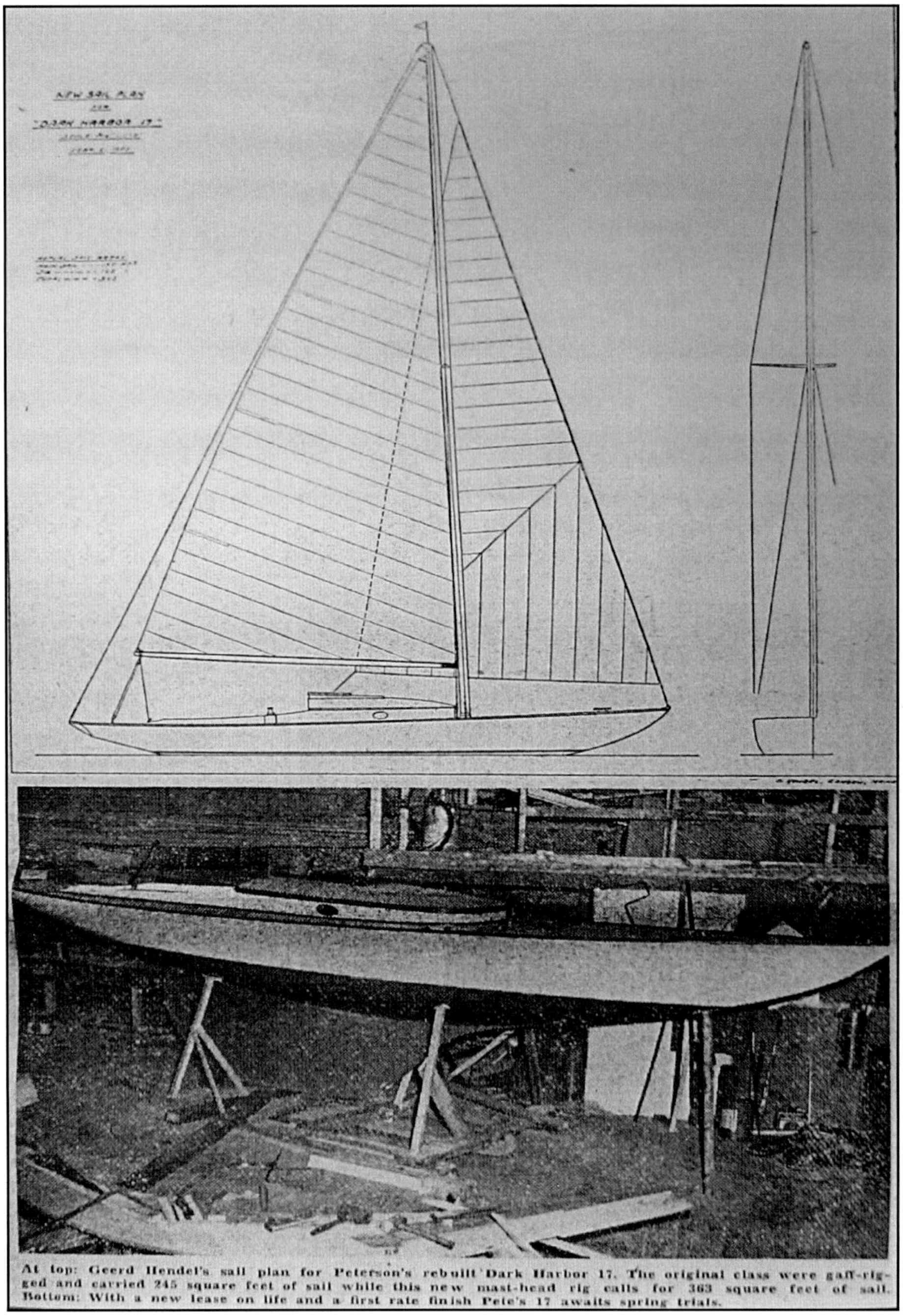

At top: Geerd Hendel's sail plan for Peterson's rebuilt Dark Harbor 17. The original class were gaff-rigged and carried 245 square feet of sail while this new mast-head rig calls for 363 square feet of sail. Bottom: With a new lease on life and a first rate finish Pete's 17 awaits spring trials.

The caption at the bottom of the image reads: "At top: Geerd Hendel's sail plan for Peterson's [William 'Pete' Peterson, owner of Camden Shipbuilding Company] rebuilt Dark Harbor 17. The original class were gaff rigged and carried 245 square feet of sail while this new mast-head rig calls for 363 square feet of sail. With a new lease on life and a first-rate finish, Pete's 17 awaits spring trials."

Maine Coast Fisherman, March 1952.

1957–59: *Widgeon* was a 26' auxiliary sloop designed for S.B. Foot and built by Norman Hodgdon. Silas B. Foot III was president of the S.B. Foot Tanning Company, which was founded just outside Red Wing, Minnesota, in 1872. He was the fourth generation of the Foot family to preside over the business and would become president prior to selling it to the Red Wing Shoe Company in 1986.[85]

Widgeon, now known as *Jackie B*, is owned by Murray Bates and named for his daughter, Jackie Bates. The boat was refitted at Sid Norton's yard on Seven Hundred Acre Island, Islesboro, Maine, about 1959.

Hendel's design for *Widgeon* was nearly identical to that of *Whistler*, although *Widgeon* was built as a wooden boat. Silas Foot wrote to Hendel in 1959 seeking guidance on how to increase her speed under sail, which was the same issue that Hendel addressed in sailing *Whistler*. In reply, Hendel advised Foot that to achieve greater speed with *Whistler*, he had installed a short bowsprit so she could carry a large low-cut jib. That jib had been replaced with a new one with a higher-cut profile, but Hendel had kept the original jib and offered it to Foot to use for testing. Hendel noted that in light winds, the low-cut jib would draw better than a higher-cut jib. He also offered Foot a used spinnaker with which to experiment.

Widgeon.
Geerd Hendel Collection, courtesy of his stepdaughter, Brigitte M. Gefrich.

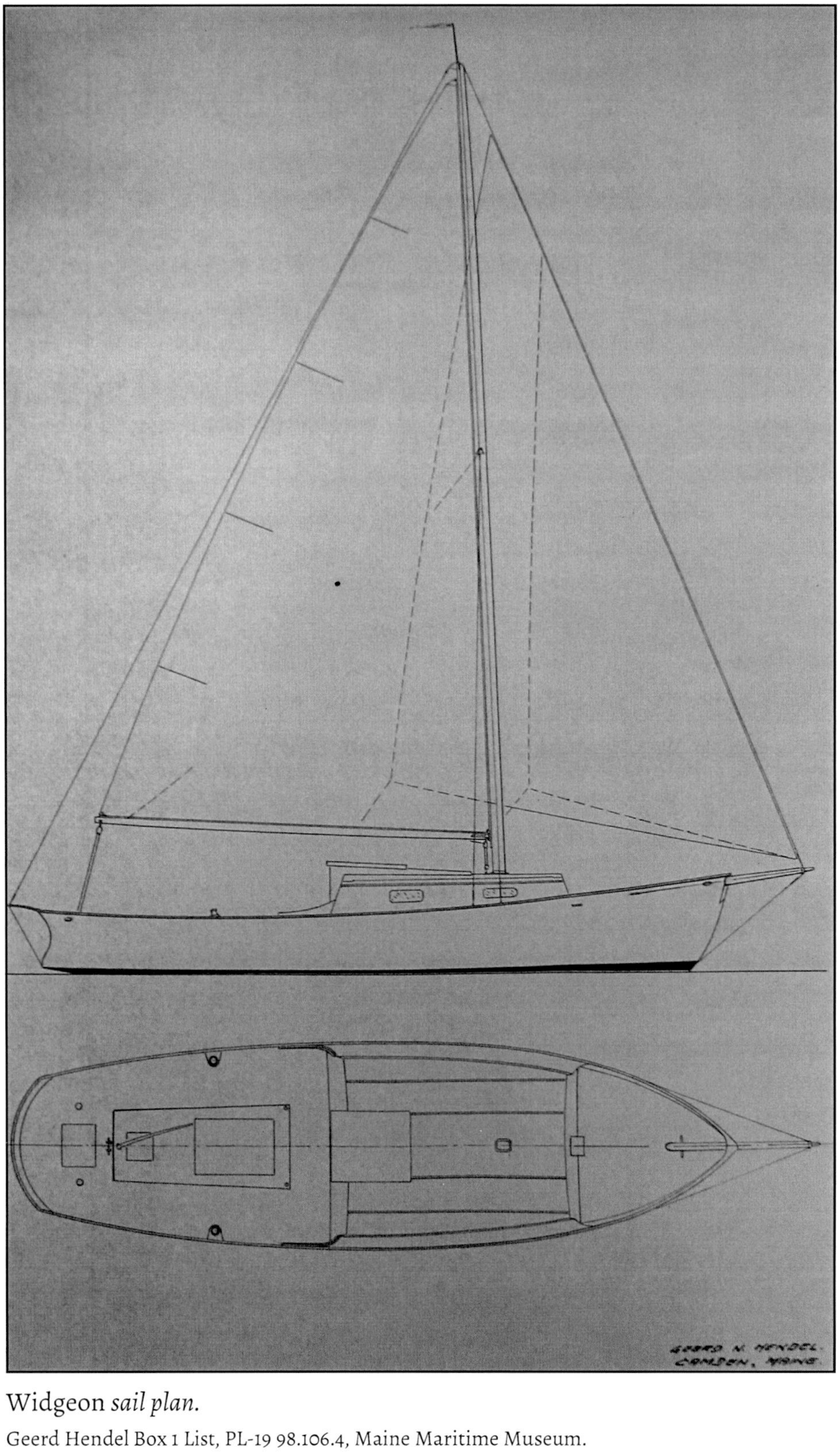

Widgeon *sail plan.*

Geerd Hendel Box 1 List, PL-19 98.106.4, Maine Maritime Museum.

Jackie B *at her Bucks Harbor, Maine, mooring.*

Maynard Bray photo, Penobscot Marine Museum.

1957: A 40' auxiliary ketch. This Hendel-designed 40' ketch was available through Camden Shipbuilding as a production boat and was capable of being sailed around the world. It featured an amidships cockpit and accommodations for six in three cabins; see the following advertisement about the vessel. The 40' ketches *The Frances* and *Freya* were built from this design (see Chapter 10).

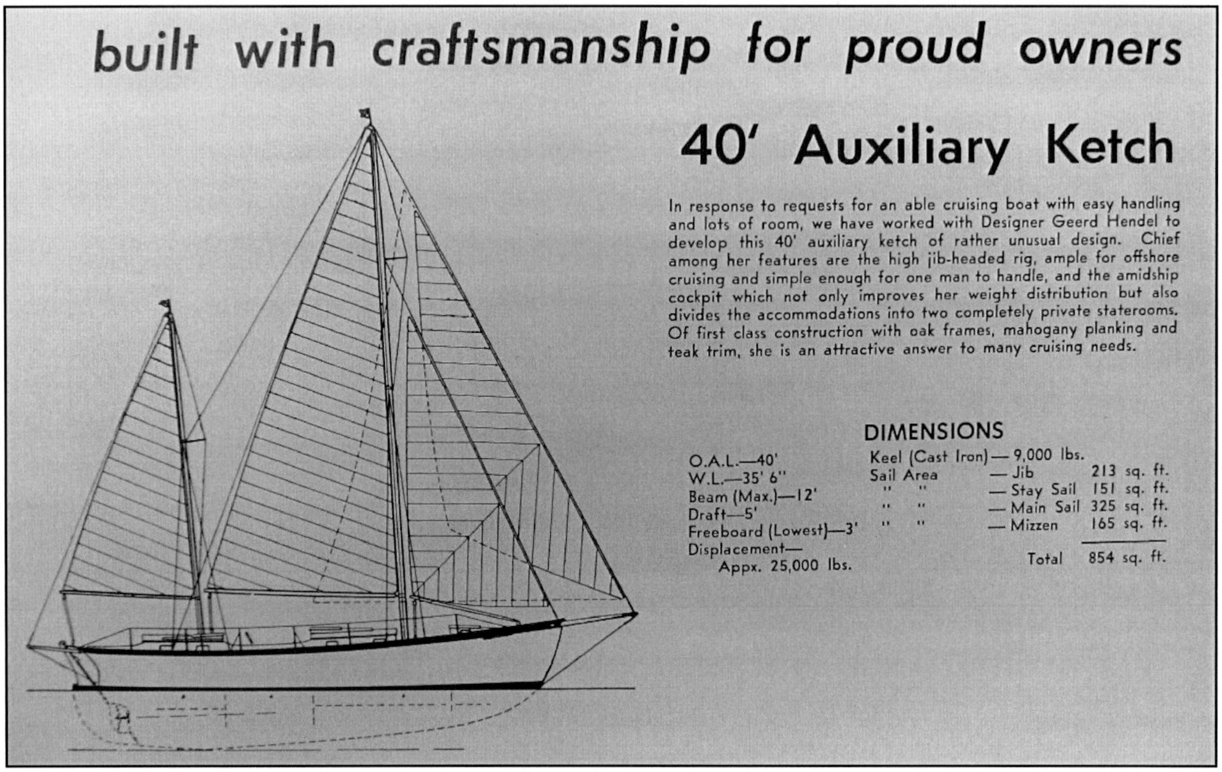

Barbara Dyer Collection, Walsh History Center, Camden Public Library.

The text of the Camden Shipbuilding ad reads as follows:

> *In response to requests for an able cruising boat with easy handling and a lot of room, we have worked with designer Geerd Hendel to develop this 40' auxiliary ketch of rather unusual design. Chief among her features are a high jib-headed rig, ample for offshore cruising and simple enough for one man to handle, and the amidship[s] cockpit, which not only improves her weight distribution but also divides the accommodations into two completely separate staterooms. Of first class construction with oak frames, mahogany planking, and teak trim, she is an attractive answer to many cruising needs.*

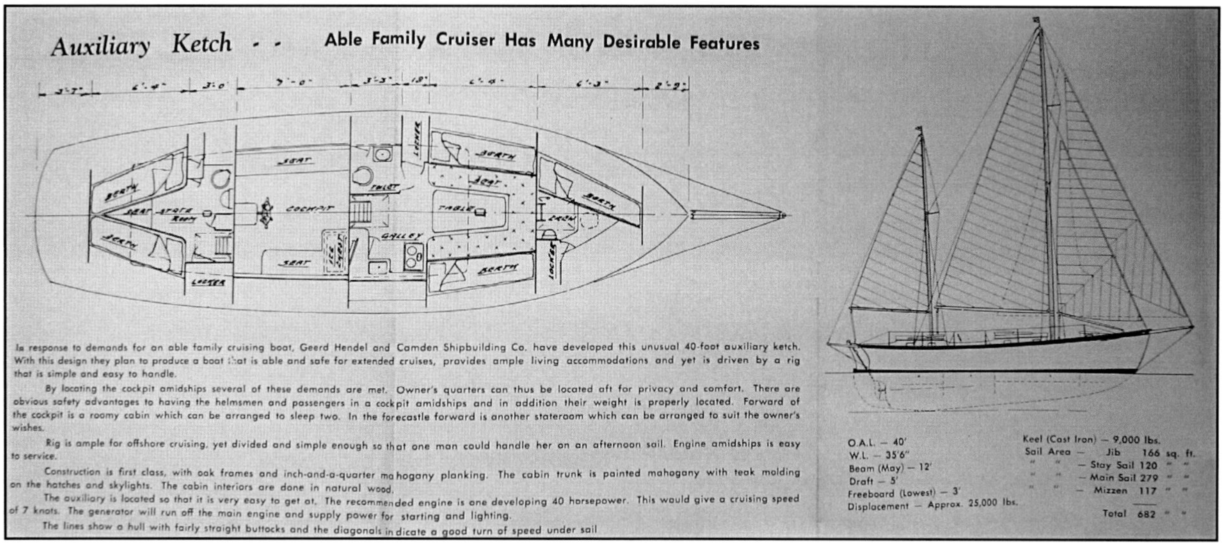

40' auxiliary ketch, detailed description.

Text from above ad:

Auxiliary Ketch...Able Family Cruiser Has Many Desirable Features

In response to demands for an able family cruising boat, Geerd Hendel and Camden Shipbuilding Co. have developed this unusual 40-foot auxiliary ketch. With this design, they plan to produce a boat that is able and safe for extended cruises, provides ample living accommodations, and yet is driven by a rig that is very simple and easy to handle.

By locating the cockpit amidships, several of these demands are met. Owner's quarters can be located for privacy and comfort. There are obvious safety advantages to having the helmsman and passengers in a cockpit amidships, and in addition, their weight is properly located. Forward of the cockpit there is a roomy cabin which can be arranged to sleep two. In the forecastle forward is another stateroom which can be arranged to suit the owner's wishes.

Rig is ample for offshore cruising, yet is divided and simple enough so that one man could handle her on an afternoon sail. Engine amidships is easy to service.

Construction is first class, with oak frames and inch-and-a-quarter mahogany planking. The cabin trunk is painted mahogany with teak molding on the hatches and skylights. The cabin interiors are done in natural wood.

The auxiliary is located so that it is very easy to get at. The recommended engine is one developing 40 horsepower. This would give a cruising speed of 7 knots. The generator will run off the main engine and supply power for starting and lighting.

The lines show a hull with fairly straight buttocks, and diagonals indicate a good turn of speed under sail.

LOA: 49'
WL: 35'6
Beam: 12'
Draft: 5'
Freeboard (lowest): 3'
Displacement: Approx. 25,000 lbs.

Keel (cast iron): 9,000 lbs.
Sail area, jib: 199 sq. ft.
Sail area, staysail: 120 sq. ft.
Sail area, mainsail: 279 sq. ft.
Sail area, mizzen: 117 sq. ft.

The Frances, *a 40' ketch built in 1962, often called simply* Frances *(later* Windsway *and* Cousin Elizabeth*).*

Barbara Dyer Collection, Walsh History Center, Camden Public Library.

The launching of *The Frances,* built for Philip E. Merryman of Trumbull, Connecticut, took place on December 27, 1962. Built at Camden Shipbuilding to the Hendel-designed plans for a 40' ketch, she was mahogany-planked over oak frames and Everdur fastened. Deck, hatches, and trim were of teak, spars were Sitka spruce, and she carried Dacron sails made by Henry Bohndell of Rockport, Maine.

The boat was powered by a GM 3-54 67-hp diesel and had a 120-volt Onan generator with a 12-volt Delco alternator.

The ketch was equipped for trans-Atlantic cruising and carried enough fuel and water for Pacific Ocean voyaging, since the owner was planning an around-the-world cruise.

Electrical and electronic equipment included an Apelco automatic pilot, a depth sounder, a radio transmitter and receiver for both voice and key, a radio marine

Hi-Seas receiver, Triton Marine automatic direction finder, Sum Log speedometer, and Parr electric water pumps (for fresh and salt water) for bilge, fire, and decks. The boat also had an electric hot water heater, range, and radiant heaters.

In 1967, a 40' ketch *Freya,* sister vessel of *The Frances,* was built at O. Lie-Nielson's "Lee's Boat Shop" on Park Street in Rockland, Maine, for Donald Parson of New York, New York, and Brooklin, Maine. It had accommodations for six.[86]

In 1961, Hendel designed Elisa Rymes, *a 32' double-headsail cutter (later* Night Heron *and* Tahra*), built by Bigelow Bros. in Monument Beach, Massachusetts, with a 10' beam, 5'9" draft, and Friendship sloop bow.*
WoodenBoat Registry.

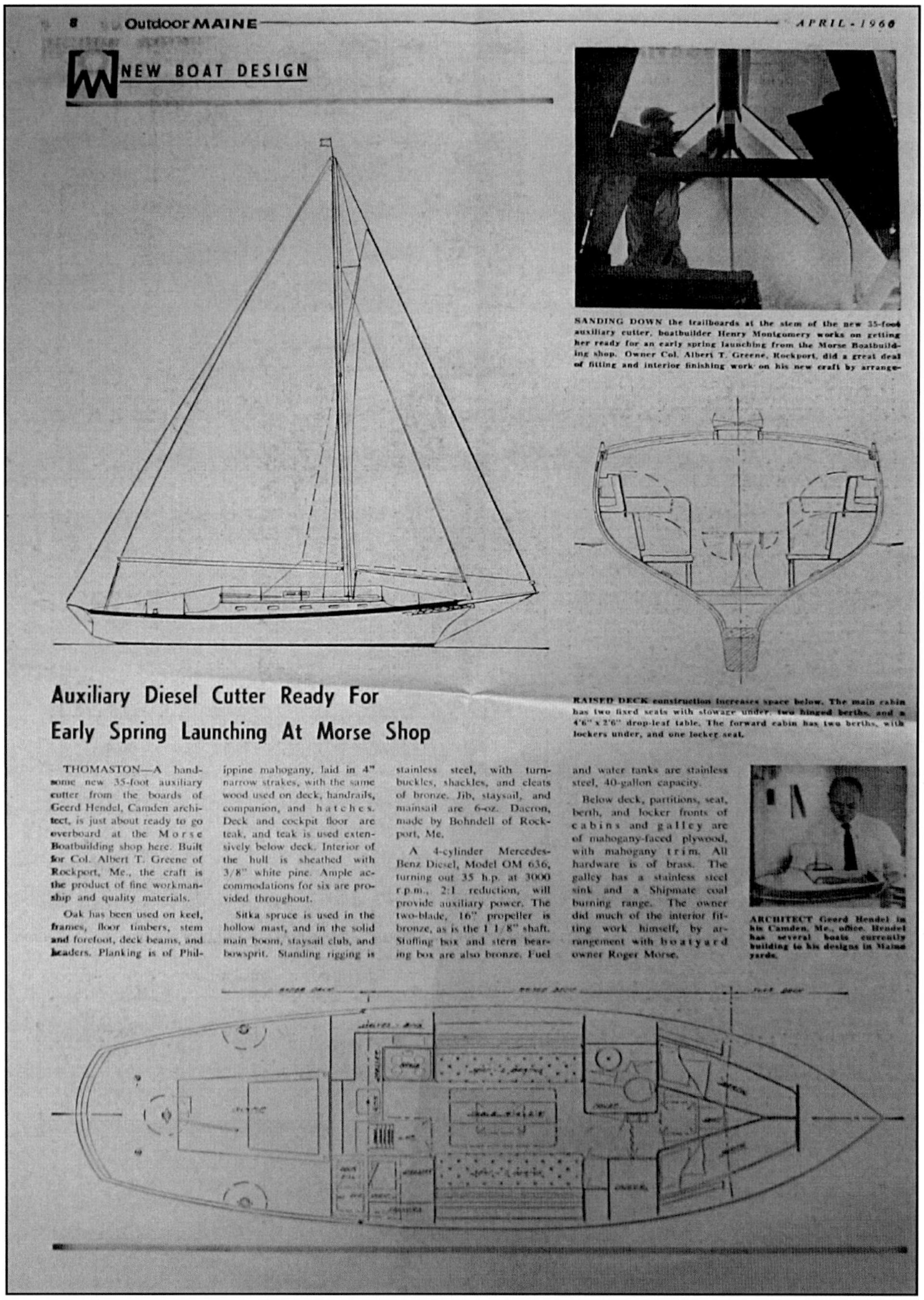

8 Outdoor MAINE — APRIL - 1966

NEW BOAT DESIGN

SANDING DOWN the trailboards at the stem of the new 35-foot auxiliary cutter, boatbuilder Henry Montgomery works on getting her ready for an early spring launching from the Morse Boatbuilding shop. Owner Col. Albert T. Greene, Rockport, did a great deal of fitting and interior finishing work on his new craft by arrange-

RAISED DECK construction increases space below. The main cabin has two fixed seats with stowage under, two hinged berths, and a 4'6" x 2'6" drop-leaf table. The forward cabin has two berths, with lockers under, and one locker seat.

Auxiliary Diesel Cutter Ready For Early Spring Launching At Morse Shop

THOMASTON—A handsome new 35-foot auxiliary cutter from the boards of Geerd Hendel, Camden architect, is just about ready to go overboard at the Morse Boatbuilding shop here. Built for Col. Albert T. Greene of Rockport, Me., the craft is the product of fine workmanship and quality materials.

Oak has been used on keel, frames, floor timbers, stem and forefoot, deck beams, and headers. Planking is of Philippine mahogany, laid in 4" narrow strakes, with the same wood used on deck, handrails, companion, and hatches. Deck and cockpit floor are teak, and teak is used extensively below deck. Interior of the hull is sheathed with 3/8" white pine. Ample accommodations for six are provided throughout.

Sitka spruce is used in the hollow mast, and in the solid main boom, staysail club, and bowsprit. Standing rigging is stainless steel, with turnbuckles, shackles, and cleats of bronze. Jib, staysail, and mainsail are 6-oz. Dacron, made by Bohndell of Rockport, Me.

A 4-cylinder Mercedes-Benz Diesel, Model OM 636, turning out 35 h.p. at 3000 r.p.m., 2:1 reduction, will provide auxiliary power. The two-blade, 16" propeller is bronze, as is the 1 1/8" shaft. Stuffing box and stern bearing box are also bronze. Fuel and water tanks are stainless steel, 40-gallon capacity.

Below deck, partitions, seat, berth, and locker fronts of cabins and galley are of mahogany-faced plywood, with mahogany trim. All hardware is of brass. The galley has a stainless steel sink and a Shipmate coal burning range. The owner did much of the interior fitting work himself, by arrangement with boatyard owner Roger Morse.

ARCHITECT Geerd Hendel in his Camden, Me., office. Hendel has several boats currently building to his designs in Maine yards.

"Lady Joan," Outdoor Maine *magazine, April 1966.*

Hendel Photograph Collection, Box MS-310, Maine Maritime Museum.

The 1966 Lady Joan (later *Lorraine* and *Starbuck*) was a 35' auxiliary cutter described in *Outdoor Maine* magazine, April 1966; see full-size text following.

THOMASTON—A handsome new 35-foot auxiliary cutter from the boards of Geerd Hendel, Camden architect, is just about ready to go overboard at the Morse Boatbuilding shop here. Built for Col. Albert T. Greene of Rockport, M[aine], the craft is the product of fine workmanship and quality materials.

Oak has been used on keel, frames, floor timbers, stem, and forefoot, deck beams, and headers. Planking is of Philippine mahogany, lain [sic] in 4" narrow strakes, with the same wood used on deck, handrails, companion, and hatches. Deck and cockpit floor are teak, and teak is used extensively below deck. Interior of the hull is sheathed with 3/8" white pine. Ample accommodations for six are provided throughout.

Sitka spruce is used in the hollow mast, and in the solid main boom, staysail club, and bowsprit. Standing rigging is stainless steel with turnbuckles, shackles, and cleats of bronze. Jib, staysail, and mainsail are 6-oz. Dacron, made by Bohndell of Rockport, M[aine].

A 4-cylinder Mercedes-Benz Diesel, model OM 636, turning out 35 hp at 3000 rpm, 2:1 reduction, will provide auxiliary power. The two-blade, 16" propeller is bronze, as is the 1-1/8" shaft. Stuffing box and stern bearing box are also bronze. Fuel and water tanks are stainless steel, 40-gallon capacity.

Below deck, partitions, seat, berth, and locker fronts of cabins and galley are mahogany-faced plywood with mahogany trim. All hardware is of brass. The galley has a stainless-steel sink and a Shipmate coal-burning range. The owner did much of the interior finish work himself, by arrangement with boatyard owner Roger Morse.

The raised-deck construction increases space below. The main cabin has two fixed seats with storage under, two hinged bunks, and a 4'6" x 2'6" drop-leaf table. The forward cabin has two berths, with lockers under, and one locker seat.

Lady Joan *under sail in Penobscot Bay.*

Geerd Hendel Collection, courtesy of his stepdaughter, Brigitte M. Gefrich.

Lady Joan Launched For Col. Greene

The trim, 35-foot auxiliary sloop Lady Joan was launched last Tuesday by the Morse Boatbuilding Co., Thomaston, for Col. Alfred T. Greene, Rockport.

The cruising sloop, designed to sleep six persons, was christened by Miss Patricia Greene, daughter of Colonel and Mrs. Greene.

The Lady Joan, designed by Geerd Hendel, Camden, is planked with Philippine mahogany over native oak frames. Her trim is of Honduras mahogany, decks and cabin top are of teak. The craft's auxil- water line, has a beam of 10

iary is a 36-h.p. Mercedes diesel, with two-to-one reduction gear.

Henry Bohndell, Rockport, sewed the 777-square feet of Dacron sail. John Larsen also of Rockport, carved and goldleafed the trailboards.

The hollow mainmast and the spars are Sitka Spruce, and stainless steel wire is used in the rigging, but fittings are bronze throughout the craft. Tanks are of stainless steel. A coal briquet Shipmate heats the galley. She's 27 feet on the feet, and draws 5 ft. 9 in.

Colonel Greene served with the Marines 23 years on the East and West Coasts and in the Pacific Theater in World War II, in Infantry, Artillery, Paratroop and Supply sections.

He is a member of the Rockport Boat Club and formerly is of Old Dominion, Alexandria, Va. where he had a power cruiser.

Greene is married to the former Joan White of New York. They also have a son, Robert, a student at University of Maine. Patricia is a senior at Rockport High School.

Lady Joan, *a 35' auxiliary cutter built from Hendel's plans at the Morse Boatbuilding Company, Thomaston, Maine.*

National Fisherman, Camden, Maine. Hendel Photograph Collection, Box MS-310, Maine Maritime Museum.

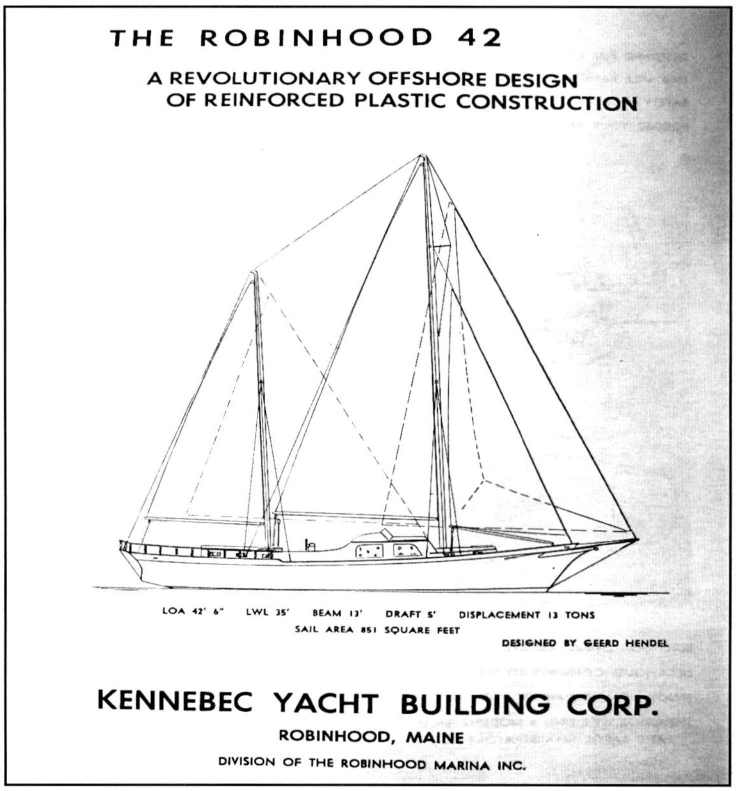

Robinhood 42 *yawl.*

Hendel Photograph Collection, Box MS-310, Maine Maritime Museum.

It is unclear from currently available records the year in which Hendel designed the Robinhood 42, which this advertisement describes as being constructed of reinforced plastic. In his 1988 interview at the Mystic Museum Research Library, Hendel observed that no vessels of this design were ever built.

In about 1956, the boatbuilding industry began to move from wood to reinforced plastic/fiberglass, which offered the prospect of building one-piece hulls that would stay watertight and allow the construction of a whole series of identical boats. The first serious production for the commercial market started in the late 1950s. Beginning in 1959, the Pearson Corp. of Bristol, Rhode Island, produced its popular 28' Triton sailboat. In 1961, Ray Greene & Co., Toledo, Ohio, started construction of what would become 175 of the 26' Sparkman & Stephens "New Horizons" sloop design.

Diana *under way with owners (l. to r.) Diana, young Eben, and Ebenezer Gay aboard.*
Diana Gay Collection.

1971: *Diana* (later *Euphoria*) is a double-headsail sloop inspired by Friendship sloop designs, with a clipper bow and wineglass stern. She was built for Ebenezer and Diana Gay of Hingham, Massachusetts, and Vinalhaven Island, Maine, by the Jarvis Newman Yard (hull P03) in Southwest Harbor. The demand for Newman's hulls became so great that he made the decision to build, with only three employees, boat hulls with deck and engine installed and ship them to other boat finishers. *Diana* was finished by Jim Rockefeller's Bald Mountain Boat Works in Camden. Hendel designed her interior layout, and may have designed her hull, as well, but source materials don't agree on that. Eben and Diana Gay owned her for 35 years (until 2006). She carries sail #94 in the Friendship Sloop Society's register, is 37' LOA, 32' length on deck (LOD), and has a beam of 10' and draft of 5'6". Her construction is carvel-planked mahogany over oak frames 9" on center; gross displacement is 17,960 lbs.; and her 6000-lb. keel is iron. She carries 684 square feet of sail on a masthead rig, has a self-tending staysail, and power is a Volvo Penta 2B marine diesel. She features standing headroom below and two settees with Pullman-style berths overhead. Under way, she was easily recognizable with her distinctive tanbark sails.

The sloop was extensively rebuilt in 1977 by boatbuilder Philip Edson, Indian Orchard, Massachusetts. Her new owner in 2006 made hull and deck improvements and added electric wiring for navigation and cabin lights.

CHAPTER NINE

Fishing Vessels, Lobster Boats, and Cruisers

Draggers and Trawlers

Hendel's knowledge of fishing draggers and trawlers was partially based on experience gained while working at BIW. As noted earlier, in 1932 BIW was awarded the construction work for 18 fishing trawlers for Francis J. O'Hara's Boston-based fishing fleet. Then in 1936, BIW secured work for building nine additional trawlers: four of 145' for Bay State Fishing Co.; two for Usen Fishing Co.; two of 136' for F.J. O'Hara & Sons; and one of 147.5'.[87]

Several photographs of draggers built at BIW in the 1930s are contained in Hendel's personal files at the Maine Maritime Museum, including that of the 1936 beam trawler *Surf.*

145' trawler Surf.

Hendel Photograph Collection, MS-310, Maine Maritime Museum.

Some sources describe draggers as being less than 100' LOA and usually wood-hulled, while those exceeding that length are called trawlers and are generally built of steel. Draggers work closer to the coastline, while trawlers are offshore fishing vessels. Each drags a large conical fishnet along the seafloor to catch cod, haddock, redfish, and flounder, and sometimes they also fish for shrimp or scallops. An "eastern-rig" dragger or trawler fishes over the side and has its working deck amidships and wheelhouse aft.
"Stern" trawlers fish over the stern and have a wheelhouse forward of the working deck.

After Hendel had started his own independent naval architecture design practice, between 1944 and 1952 he designed at least seven wooden-hulled draggers in the 65' to 90' range, five of which were built at the Harvey Gamage yard in South Bristol, Maine.

The 65' dragger *Camden* was designed by Hendel and built in 1944 at Camden Shipbuilding; the photograph shows its homeport to be New Bedford, Massachusetts.

In the post–World War II period between 1945 and 1952, the four Hendel-designed eastern rig draggers *Althea Joyce, Wild Duck, The Clipper Fish,* and *Albatross* were built by the Harvey Gamage yard in South Bristol, Maine, and have been documented by the South Bristol Historical Society.

1945: *Althea Joyce,* a 61' wooden dragger (hull 137). Her original owner was Captain Salvatore.

Dragger Camden *on the ways at Camden Shipbuilding.*
Barbara Dyer Collection, Walsh History Center, Camden Public Library.

The three 90' draggers built in 1946 (*Albatross, The Clipper Fish,* and *Wild Duck*) appear to be sister ships constructed from the same plans and specifications.

1946: Wild Duck, *a 90' wooden dragger with a beam of 21.9', an 11.8' draft, 132 gross tons and 89 net tons, with a 300–400-hp diesel as power (hull 144).*

Hendel Photograph Collection, Box 3, Folder 12, Maine Maritime Museum.

1946: The Clipper Fish *(later* Clipper*), a 90' "eastern-rig" wood-hulled dragger with a 21.9' beam and 11.8' draft, launched November 30, 1946 (hull 146). She was 132 gross tons and 89 net tons with a 300–400-hp diesel as power. The original owner was A. Arruda.*

Hendel Photograph Collection, Box 1, Folder 18, Maine Maritime Museum.

The Clipper Fish*'s stern shows her homeport as New Bedford, Massachusetts.*
Geerd Hendel Collection, courtesy of his stepdaughter, Brigitte M. Gefrich.

1946: Albatross*, a 90' wooden dragger with a beam of 21.9' and draft of 11.8' (hull 147), was built at the Harvey Gamage Shipyard. Gross tonnage was 132, net tonnage 89, and power was a 300–400-hp diesel. The South Bristol Historical Society's file notes say the vessel was rammed and sunk when new, but she apparently was raised and returned to service.*
Hendel Photograph Collection, Box 1, Folder 3, Maine Maritime Museum.

A model of *Albatross* was built by boat model maker John "Jay" S. Hanna of Rockport, Maine, for the Smithsonian's National Museum of American History archives, Maritime History section, Washington, DC.

John "Jay" S. Hanna's model of the fishing trawler Albatross. *Hanna estimated that 850 hours of work were needed to create the model.*

Smithsonian Institution Archives, Museum of American History, Maritime History section.

1946: *Pam-Ann*, a 99'8" dragger built under the supervision of the American Bureau of Shipping (ABS) by the Casey Boat Building Company, Inc., of Fairhaven, Massachusetts. The vessel's owner was Capt. Russell Grinnell Jr.

Hendel's correspondence files at the Maine Maritime Museum make it clear that Rice Brothers Corp. of East Boothbay was also awarded a contract to build a 100' dragger of his design by Capt. Russell Grinnell Jr., with construction to start in early 1946.

In 1951, Hendel prepared a design plan for a 50' shrimp trawler, but no file information is provided concerning his potential client.

Dragger Pam-Ann.

Atlantic Fisherman, January 1951, Hendel Collection, Maine Maritime Museum.

Hendel developed this design proposal for a 60' or 64' sardine carrier in 1947.
Hendel Collection, Box 1, List #6, PL-19 98.106.4, Maine Maritime Museum.

1952: Pocahontas *(later the* Michelle *and* Annette*) launched at the Harvey Gamage Shipyard (hull 176) on February 3, 1952. She was a wood-hulled scallop dragger 92' LOA and 130 tons, powered by a Caterpillar D397 diesel, and owned by the Pocahontas Fishing Co., Rockland, Maine.*
David Andrews Collection.

Pocahontas *model by John "Jay" S. Hanna, Rockport, Maine.*
Maine Maritime Museum Collection.

Pocahontas *model by John "Jay" S. Hanna, Rockport, Maine.*
Maine Maritime Museum Collection.

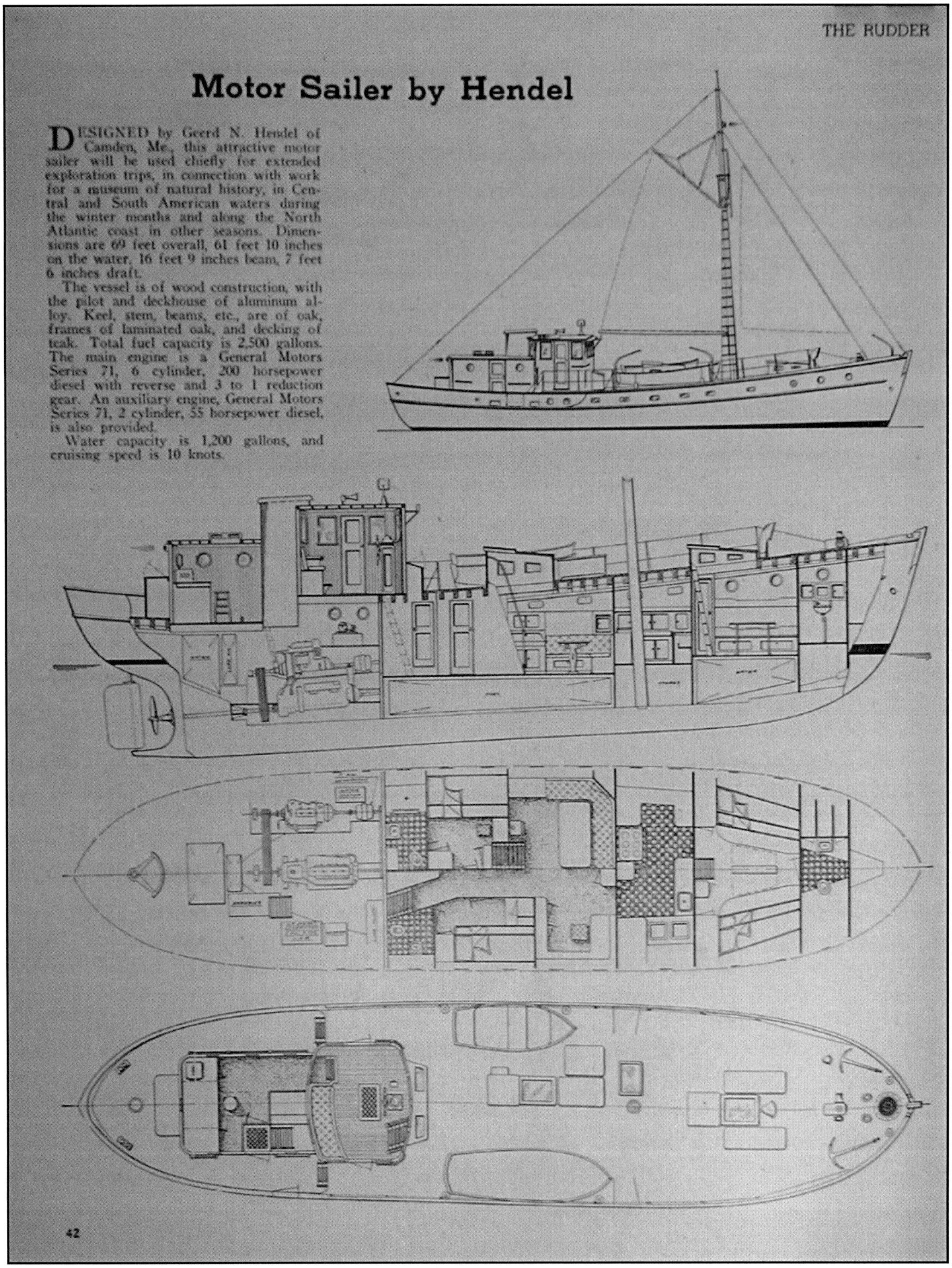

THE RUDDER

Motor Sailer by Hendel

DESIGNED by Geerd N. Hendel of Camden, Me., this attractive motor sailer will be used chiefly for extended exploration trips, in connection with work for a museum of natural history, in Central and South American waters during the winter months and along the North Atlantic coast in other seasons. Dimensions are 69 feet overall, 61 feet 10 inches on the water, 16 feet 9 inches beam, 7 feet 6 inches draft.

The vessel is of wood construction, with the pilot and deckhouse of aluminum alloy. Keel, stem, beams, etc., are of oak, frames of laminated oak, and decking of teak. Total fuel capacity is 2,500 gallons. The main engine is a General Motors Series 71, 6 cylinder, 200 horsepower diesel with reverse and 3 to 1 reduction gear. An auxiliary engine, General Motors Series 71, 2 cylinder, 55 horsepower diesel, is also provided.

Water capacity is 1,200 gallons, and cruising speed is 10 knots.

42

1948: This 69' motor sailer for an (unnamed) museum of natural history for use in Central and South American waters, had an aluminum-alloy pilothouse and deckhouse.

Hendel Collection, *The Rudder* magazine, April 1948. Maine Maritime Museum.

The motor sailer text reads:

> *DESIGNED by Geerd N. Hendel of Camden, M[aine], this attractive motor sailer will be used chiefly for extended exploration trips in connection with work for a museum of natural history, in Central and South American waters during the winter months and along the North American coasts in other seasons. Dimensions are 69 feet overall, 61 feet 10 inches on the water, 16 feet 9 inches beam, and 7 feet 6 inches draft.*
>
> *The vessel is of wood construction with the pilot and deckhouse of aluminum alloy. Keel, stem, beams, etc., are of oak, frames of laminated oak, and decking of teak. Total fuel capacity is 2,500 gallons. The main engine is a General Motors Series 71, 6-cylinder, 200-horsepower diesel with reverse and a 3-to-1 reduction gear. An auxiliary engine, General Motors Series 71, 2-cylinder 55-horsepower diesel, is also provided.*
>
> *Water capacity is 1,200 gallons, and cruising speed is 10 knots.*

There is no documentation, however, that confirms this vessel was ever built.

1956: Bingo III, *a 43' fishing and passenger boat built by Axel Gronros's Rockland Boat Shop in Rockland, Maine, for Knute Lovgren of Brielle, New Jersey.*
Mel Deak Collection.

Sunbeam III (1939–1964) was owned by the Maine Seacoast Mission. Founded in 1905, the Maine Seacoast Mission offers youth programs, telemedicine, housing rehabilitation, food pantries, and college scholarships to island communities. *Sunbeam III* was designed by Tams Inc., and built in 1939 by the Harry G. Marr Shipyard in Damariscotta, Maine (previously the Jonah D. Morse yard). Hendel referenced this vessel in his 1988 audio interview and had a major design role in her pilothouse refit at Camden Shipbuilding. The Maine Seacoast Mission continues to function to this day with their current vessel, *Sunbeam V*.

Sunbeam III.
Barbara Dyer Collection, Walsh History Center, Camden Public Library.

In 1959, Hendel prepared design plans for Rupert Neily Jr., President of Maine Coast Boat Sales, for the construction of a boat basin to be called the "Boatel Marina" at 80 Commercial Street in Boothbay Harbor. It included 500' of float space, could accommodate vessels of up to 80' in length, and was the only marina project that he designed. The site is now the location of the Tugboat Inn and Marina.

Page from an undated Camden Shipbuilding marketing brochure showing Sunbeam III *hauled-out on their marine railway.*

Barbara Dyer Collection, Walsh History Center, Camden Public Library.

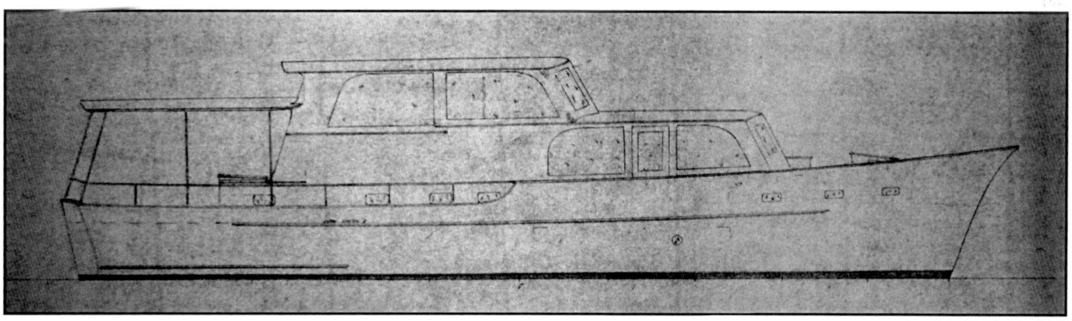

Launched in 1960, La Reine *(later* Driftwood II *and* Four Walls II*), a power cruiser built by the Newbert & Wallace Shipyard in Thomaston for Carlisle V. Watson of Portland and Camden, Maine. She was a wooden-hulled recreational vessel, 53' LOA with a beam of 13.3' and draft of 7.1'. She was based in Florida for some years, but her trail has since disappeared.*

Geerd Hendel Box #1, List #6, PL-19 98.106.4, Maine Maritime Museum.

1960: The Rockhopper *(later* Seven Brothers*) built at the Newbert & Wallace yard, a 33' lobster/pleasure boat for Dr. David C. Dewees of Wakefield, Rhode Island. Dr. Dewees added a 16' bowsprit and tower to the vessel for sword fishing.*
Rodney Cook Collection.

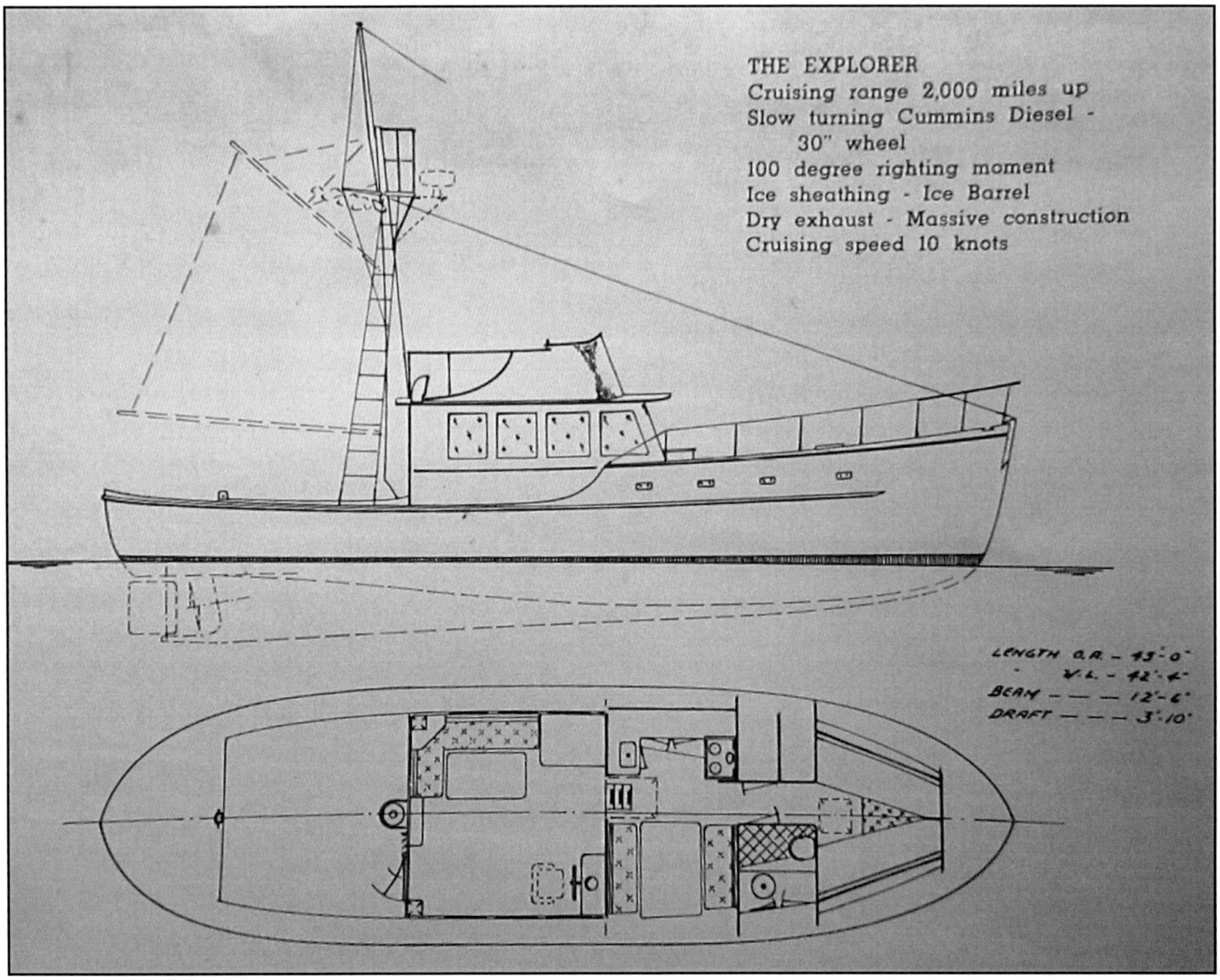

1963 design for the 43' cruiser Explorer design prepared for Bald Mountain Boat Works. There are no records documenting that any vessels of this specific double-ended design were ever built. Hendel's design for 1970 Kyloe*, with 45' LOA, however, nearly duplicates the Explorer's detail.*
Hendel Photograph Collection, MS-310, Maine Maritime Museum.

Hendel's 1960s plan profile for a Down East-34 hull.

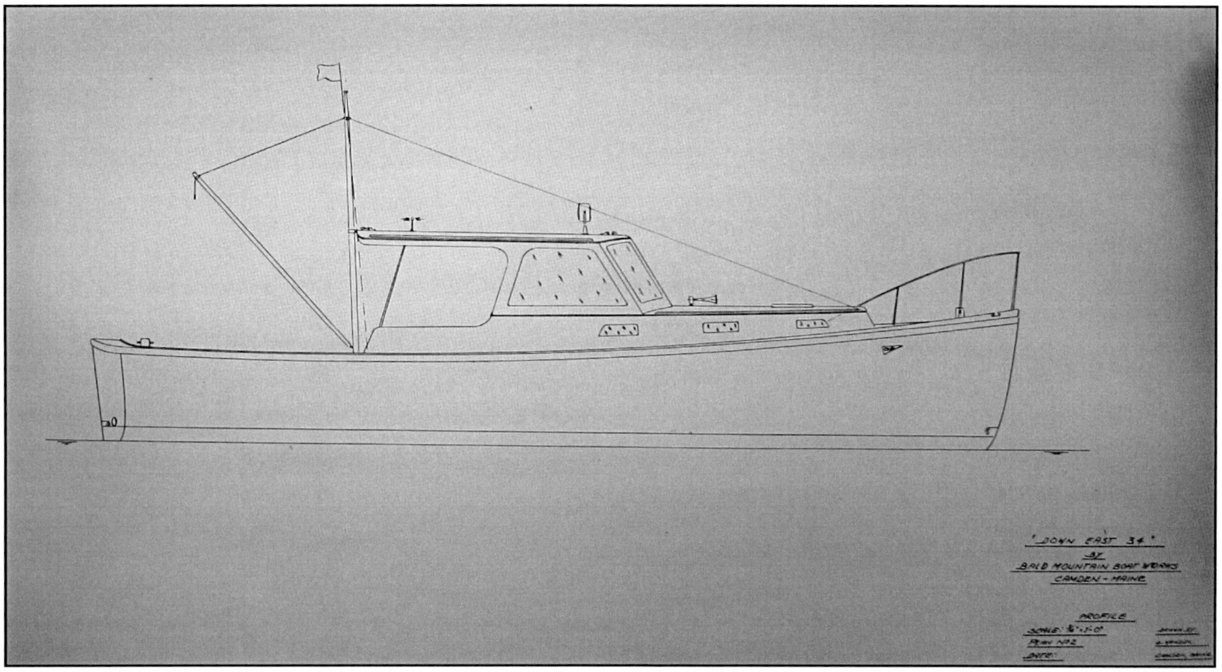

Down East 34.

Hendel Photograph Collection, File 1, Folder 19, Maine Maritime Museum.

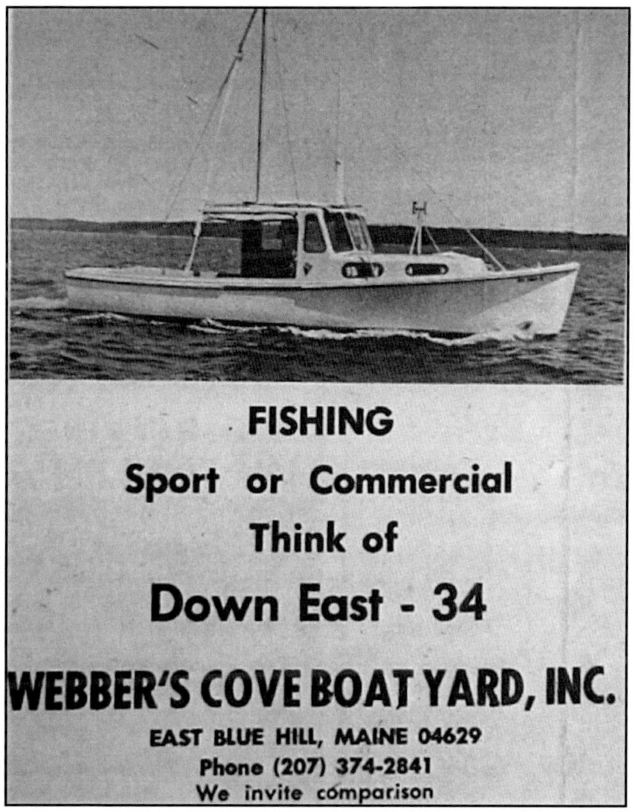

Mid-1960s: Hendel designed and invoiced Bald Mountain Boat Works for a profile and arrangement plan for a Down East 34 fiberglass hull from the Webber's Cove Boat Yard (WEBCO) in Blue Hill, but there's no record that the boat was ever built.

Hendel Photograph Collection, MS-310, Maine Maritime Museum.

1965: *Brennain* (later *Lone Star*) a 27' sterndrive cruiser built with a production hull at PADEBCO Custom Boats in Round Pond, Maine. She was owned by Jack and Margaret Neeson of Spruce Head Island, Maine. Readers of *The Skipper* magazine might recall that Margaret had written its monthly "The Ditty Bag" column for almost 20 years. Hendel designed the cabin and its interior arrangement; the cruiser had a completely equipped galley with a Shipmate stove and could sleep four.

Their son John D. "JD" Neeson said the boat cost $6,400 when new. He also noted that Hendel's design for the cabin structure called for the cabin sole to be dropped by 8" to lower the cabin profile. From Hendel's aesthetic perspective, this change significantly improved the proportional relationship between the cabin and hull, and it had the added benefit of creating more headroom.

Brennain.

Hendel Photograph Collection, MS-310, Box 1, Maine Maritime Museum.

1966: a 27' pleasure, utility, or lobster boat. This copy of an advertisement in the *National Fisherman* for this vessel was contained in the Hendel files at the Maine Maritime Museum.

In 1965, the first 27-footer of this design was built by A.T. Greene as a floating laboratory for Michigan State University's W.K. Kellogg Biological Station on Gull Lake. The Station serves to increase peoples' understanding of natural and managed ecosystems and their linkages to society.

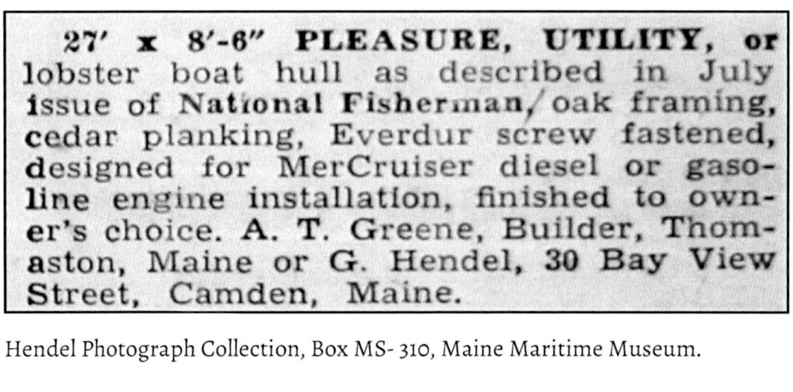

27' x 8'-6" PLEASURE, UTILITY, or lobster boat hull as described in July issue of **National Fisherman**/ oak framing, cedar planking, Everdur screw fastened, designed for MerCruiser diesel or gasoline engine installation, finished to owner's choice. A. T. Greene, Builder, Thomaston, Maine or G. Hendel, 30 Bay View Street, Camden, Maine.

Hendel Photograph Collection, Box MS- 310, Maine Maritime Museum.

Vicil Has Modified Maine Lobsterboat Hull.

Day Cruiser Has Lobsterboat Lines

THOMASTON, ME. — The 27′ pleasure craft Vicil was launched from the Alfred T. Greene boatshop on June 20. Vicil was sponsored by Mrs. Viola Sargent, wife of owner Philip A. Sargent of Burnt Island, Sprucehead, Me.

Sargent will use the boat primarily as a day boat, although she has two berths forward, enclosed head, compact galley with stainless sink and two-burner, Sea Cook alcohol stove and ample cupboards.

The boat was designed by Geerd N. Hendel, N. A., of Camden, Me., with a modified Maine lobsterboat hull readily adaptable to several designs of cabin and superstructure. The prototype was built for Michigan State University for use on Gull Lake as a floating research laboratory. Vicil is the second boat built by Greene to this design.

The trim little craft is 27′ x 8′6″ with 28″ draft. Built of native cedar on oak, with Everdur fastenings, she is powered by a 160 h.p. gasoline Mercruiser sterndrive. All windows are ¼″ safety plate glass.

Vicil *was the second vessel of the 27′ design, built about 1966 by A.T. Greene for Philip A. Sargent of Round Pond, Maine, to be used as a day cruiser.*

Hendel Photograph Collection, MS-310, Maine Maritime Museum.

1968: 27' Wandel, *a cruiser built at Sam Guild boat shop, Thomaston, Maine, for Dr. E.C. Kunkle of South Windham, Maine. Power was a Mercruiser 120-hp sterndrive.*

Hendel Photograph Collection,
MS-Box 310, Maine Maritime Museum.

34' lobster boat built by the Reinforced Plastics Co., Gouldsboro, Maine.

Hendel Photograph Collection, Box 1, Folder 36, Maine Maritime Museum.

July 1968 *National Fisherman* article by Irving Nevells:

> HEADING FOR MATINICUS ISLAND off the coast of Maine is the open-type lobster boat Evie Sue. The 30' x 9.6' x 2.9' boat was built by Gerry Libby of Melvin Heights, Camden, M[aine] for owner Crosby Ames of Matinicus. She was designed by Geerd Hendel, N.A. of Camden, and christened by Evelyn Ames, 15, daughter of the owner. The boat is cedar planked with oak frames and fastened with galvanized clinch nails. Power is a 165-hp gas engine.

1968: 30' Evie Sue, *an open lobster boat, built by Gerry Libby of the Melvin Heights Boat Shop, Camden, Maine.*
Irving Nevells Collection, Penobscot Marine Museum.

1969: 38' Phalarope, *a combination pilot/lobster boat, built by the Melvin Heights Boat Shop, Camden, Maine.*
Irving Nevells Collection, Penobscot Marine Museum.

From the *Courier-Gazette*, Rockland, Maine:

> A 38' COMBINATION fishing and pilot boat, the *Phalarope* is shown during her launching in a driving rainstorm. The handsome craft will see service in the ocean waters off Monhegan Island.
>
> Gerry Libby's Melvin Heights boat shop launched the 38' lobster and pilot boat *Phalarope* on August 4 from Knight's Marine Service in Rockland.
>
> A fine model of a Maine workboat, she was designed by Geerd Hendel, N.A., of Camden for Sherman Stanley of Monhegan, who will use her both for fishing and as a pilot boat for the Penbay pilots taking ships in and out of Penobscot Bay.
>
> Because of her dual role, *Phalarope* has an unusually large amount of electronic gear aboard including three radiotelephones and a Kelvin Hughes radar and depth finder.
>
> In line with late trends in lob-

ster boat construction, she has Wagner hydraulic steering and a hydraulic pot hauler from Marine Hydraulic Engineering Co. Inc. of Warren, M[aine].

The boat has 38' length, 11' 6" beam. She is of conventional wood construction with oak frame timbers 1½" x 2" on 9" centers. Keel is oak, 6" x 13". Planking is mahogany with Everdur fastenings.

The main deck and cabin roof are marine plywood covered with fiberglass, while the cockpit deck is cedar.

The engine is a GMC Toro-Flow diesel developing 220 hp at 2800 rpm. This turns a 23" x 21" three-blade wheel through a 2:1 reduction gear. The exhaust pipe goes through a hollow aluminum mast supplied by R & R Engineering of Rockland, which also furnished the two 100-gal. fuel tanks.

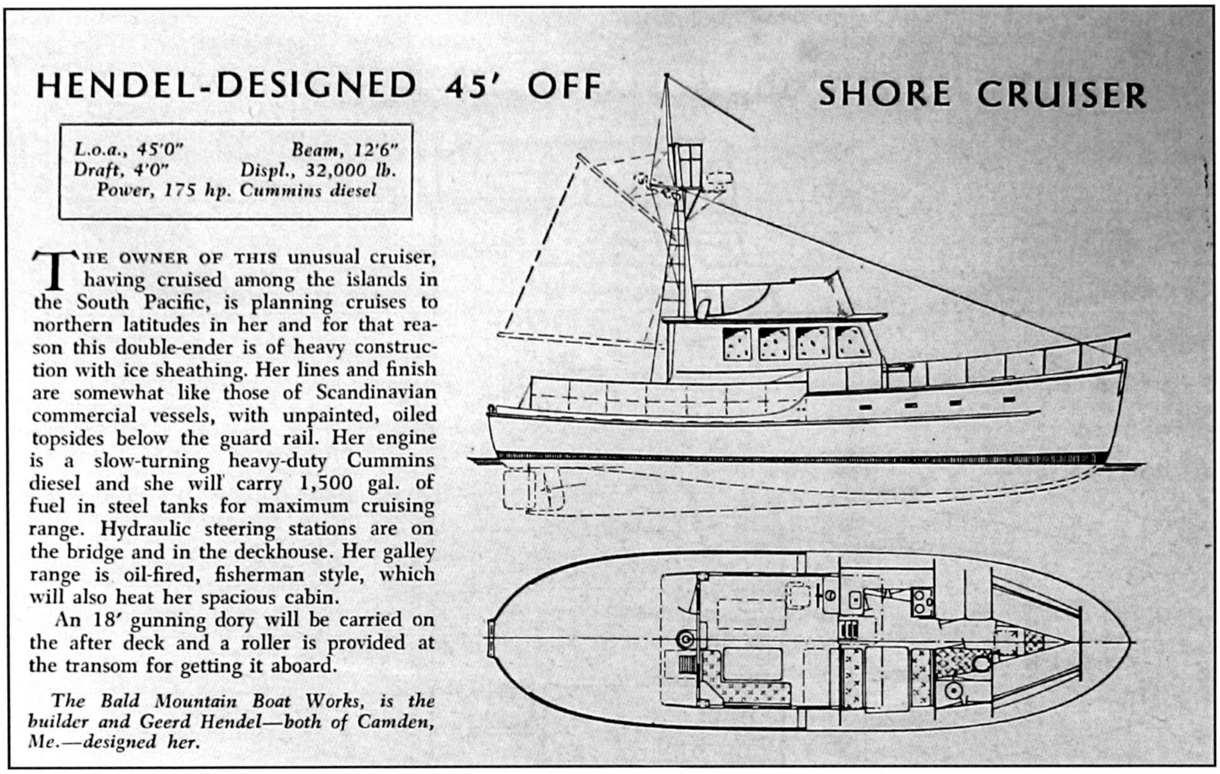

HENDEL-DESIGNED 45' OFFSHORE CRUISER

L.o.a., 45'0" *Beam, 12'6"*
Draft, 4'0" *Displ., 32,000 lb.*
Power, 175 hp. Cummins diesel

THE OWNER OF THIS unusual cruiser, having cruised among the islands in the South Pacific, is planning cruises to northern latitudes in her and for that reason this double-ender is of heavy construction with ice sheathing. Her lines and finish are somewhat like those of Scandinavian commercial vessels, with unpainted, oiled topsides below the guard rail. Her engine is a slow-turning heavy-duty Cummins diesel and she will carry 1,500 gal. of fuel in steel tanks for maximum cruising range. Hydraulic steering stations are on the bridge and in the deckhouse. Her galley range is oil-fired, fisherman style, which will also heat her spacious cabin.

An 18' gunning dory will be carried on the after deck and a roller is provided at the transom for getting it aboard.

The Bald Mountain Boat Works, is the builder and Geerd Hendel—both of Camden, Me.—designed her.

1970: Kyloe, a 45' offshore cruiser built from Hendel's plans by the Bald Mountain Boat Works.

Hendel Photograph Collection, June 1968 ad from *The Rudder* magazine, File 1, Folder 31, Maine Maritime Museum.

Moving 45′ Cruiser Down Mountain Is No Job For Two Yokes Of Oxen

By Irving Nevells

CAMDEN, ME. — Bald Mountain Boat shop, owned and operated by James S. Rockefeller Jr., is located high on the side of the mountain.

The shop started in the Rockefeller barn with the first boat being the 25′ Friendship sloop, Old Baldy. Rockefeller made quite a stir when he hauled the sloop down the mountain behind two yokes of oxen as in the days of Wilbur Morse.

It was a far cry from the little sloop to the 45′ cabin cruiser, Kyloe, which went down the mountain in September on a low-bed trailer with truck before and another behind braking on the steep grade down to the main road. Kyloe was transported to Rockland and launched by Travelift at Knight's Marine Service. Horatio Knight estimated her weight at over 25 tons with water and fuel tanks full.

Kyloe is 45′ l.o.a. with a 12′6″ beam and 4′ draft. Keel and frame are oak and planking Alaskan cedar with Everdur fastenings.

She has a 175 h.p. Cummins diesel swinging a 30″ x 21″ four-blade wheel. Top speed is about 12 knots. Kyloe was designed by Geerd Hendel, N.A., of Camden.

Owner Edward R. Noyes Jr. will operate her out of Portsmouth, N.H., and plans to live aboard most of the time. She has comfortable living quarters for six and the galley equipment includes an oil burning heater similar to that used in many kitchens ashore.

EASING 25 TONS OR MORE OF BOAT down a mountainside calls for good brakes, strong gears and a stout heart, all of which are apparently present in this view of the 45′ Kyloe moving down the driveway from James S. Rockefeller Jr.'s Bald Mountain Boat Shop. The builder follows along with camera at right.

INTERESTING LINES of the big cruiser can be seen in this photo of Kyloe shortly before her launching in Rockland, Me. Her flat run aft not only gives good bearing when she's underway, but also allows for the big propeller designed to take full advantage of her slow-turning engine. —Photos by Irving Nevells

Hendel Photograph Collection, *National Fisherman*, December 1970, Box MS-310, Maine Maritime Museum.

See enlarged text below the photographs shown above.

(Left photo) *EASING 25 TONS OR MORE OF BOAT down a mountainside calls for good brakes, strong gears, and a stout heart, all of which are apparently present in this view of the 45'* Kyloe *moving down the driveway from James S. Rockefeller Jr.'s Bald Mountain Boat Shop. The builder follows along with camera at right.*

(Right photo) *INTERESTING LINES of the big cruiser can be seen in the photo of* Kyloe *shortly before her launching in Rockland, M[aine]. Her flat run aft not only gives good bearing when she's under way, but also allows for the big propeller designed to take advantage of her slow-turning engine.*

Text above the photographs on page 121:

MOVING 45' CRUISER DOWN MOUNTAIN IS NO JOB FOR TWO YOKES OF OXEN
by Irving Nevells in the *Maine Coast Fisherman*

CAMDEN, ME—Bald Mountain Boat shop, owned and operated by James S. Rockefeller Jr., is located high on the side of the mountain.

The shop started in the Rockefeller barn with the first boat being the 25' Friendship sloop Old Baldy. *Rockefeller made quite a stir when he hauled the sloop down the mountain behind two yokes of oxen as in the days of Wilbur Morse.*

It was a far cry from the little sloop to the 45' cabin cruiser Kyloe, *which went down the mountain in September on a low-bed trailer with a truck before and another behind braking on the steep grade down to the main road.* Kyloe *was transported to Rockland and launched by Travelift at Knight's Marine Service. Horatio [Knight] estimated her weight at over 25 tons with water and fuel tanks full.*

Kyloe *is 45' l.o.a. with a 12'6" beam and 4' draft. Keel and frame are oak and planking Alaskan cedar with Everdur fastenings.*

Kyloe *under way.*
James Rockefeller Jr. Collection.

She has a 175-hp Cummins diesel swinging a 30" x 21" four-blade wheel. Top speed is about 12 knots. Kyloe *was designed by Geerd Hendel, N.A., of Camden.*

Owner Edward R. Noyes Jr. will operate her out of Portsmouth, NH, and plans to live aboard most of the time. She has comfortable living quarters for six and the galley equipment includes an oil-burning heater similar to that used in many kitchens ashore.

Bald Mountain Boat Works owner James "Jim" Rockefeller Jr. noted that the designed height of *Kyloe*'s cabin had to be increased due to the height of the owner, an alteration that caused some distortion of Hendel's original cabin/hull proportions for the vessel.

1970: Alster (pronounced "All-ster"), Hendel's own personal 33' Sportster, was built at A.T. Greene's Old Kiln Boat Works, Inc., in Rockport, Maine, and launched at Wayfarer Marine, Inc., Camden, Maine. (See Chapter 10 for Hendel's Sportster design plan and narrative, and descriptions of other Sportsters that were built.) Hendel named the boat after the Alster, a tributary that joins the Elbe in Hamburg, Germany. The Alster basin dam in the heart of Hamburg creates a shallow lake there.

Alster.
Irving Nevells Collection, Penobscot Marine Museum.

Alster, *Hendel's 33', sportster shown in 1983 with a modification to enclose part of its open cockpit.*
Hendel Photograph Collection, MS-310, Maine Maritime Museum.

The *Courier-Gazette* newspaper text by Irving Nevells for this photo reads as follows:

Naval Architect Geerd Hendel Takes Delivery of 33' Greene-Built Cruiser

CAMDEN, ME—Designing all types of craft from rowboats to ships is all in a day's work for a naval architect. When he designs a boat for his own use, that makes news.

Geerd Hendel, N.A., of Camden did it. His new boat is the 33' sport cabin cruiser *Alster*, launched June 29 at Wayfarer Marine Corp. in Camden. *Alster* was built by Col. Alfred T. Greene in the Old Kiln Boat Works Inc., in West Rockport.

This new shop, built from the ground up last year, is miles from any launching water, but this is no handicap today with transportation by lowbed trailer readily available. Several shops in this area are locating inland as waterfront sites, if available, command a premium price. There are cranes and Travelifts in Camden, Rockport, and Rockland for lowering the boats into the harbors.

Hendel's cruiser, a little large for the mobile lifts, was launched by crane.

The sport cruiser is 33' l.o.a. by 9'6" beam and draws 3'. She has oak frames and keel with mahogany planking. Fastenings are Everdur. Power is a 225-hp Palmer engine swinging a 20" x 17" three-blade Columbian wheel through a 2:1 Paragon reduction gear.

Engine controls are in the cabin and on the flying bridge. Stainless-steel tanks are by Robert Hersom, Rockport, M[aine].

There are two 60-gal. fuel tanks and one water tank holding about 25 gals.

Design speed is 15 knots.

This boat will be used mostly for day sailing in Penobscot Bay, but she has sleeping accommodations for two, one head, and [a] Shipmate propane gas stove.

1971: Blue Pigeon, *a Dyercraft 40 that received a customized cabin layout and design by Hendel, a "well known and respected naval architect," for her use as a "commuter" vessel on Penobscot Bay for Richard Krementz of Morristown, New Jersey. Krementz summered on Maine's Criehaven Island in Penobscot Bay and was a member of the Camden Yacht Club.*

Hendel Photograph Collection, Box MS-310, Maine Maritime Museum.

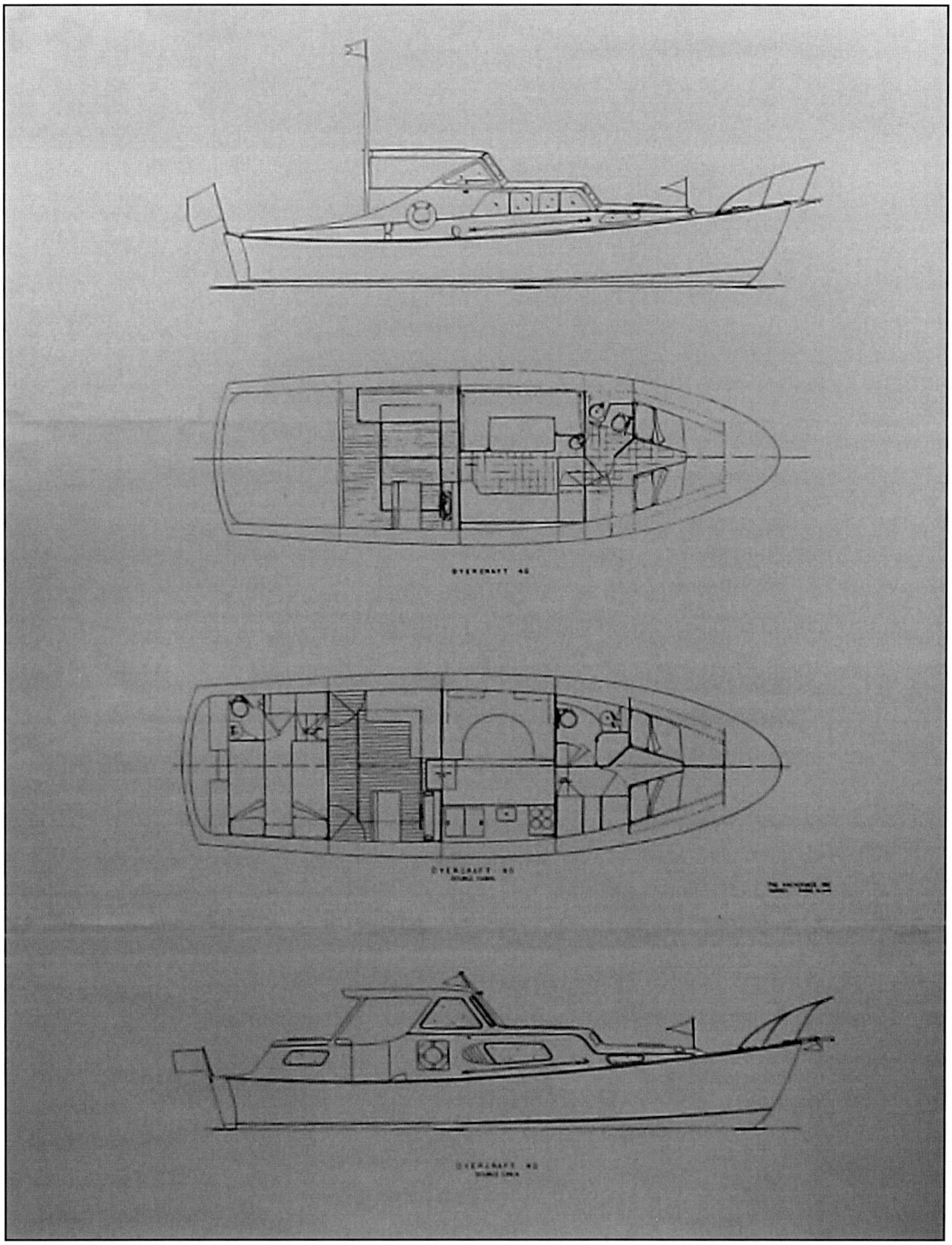

Two of Hendel's design studies for the Dyercraft 40 Blue Pigeon *customization project. Modifications included redesigning the deckhouse and bridge, adding stiffeners to the cabin structure, adding a water tank forward, and installing a ladder for diving.*

Hendel Photograph Collection, Box MS-310, Maine Maritime Museum.

CHAPTER TEN

Designs for Camden Shipbuilding & Marine Railway Company/ Camden Shipbuilding Company, Inc.

The Camden Shipbuilding & Marine Railway Company was a new corporation that began operations on January 1, 1941. Its location on Camden Harbor had been the site of boatbuilding operations since the late 1790s and was the former site of the historic Holly Bean and Robert Bean Yards. The new company's officers were Richard Lyman, president; Cary Bok, vice president; Clinton Lunt, treasurer; and attorney Parker Marean as secretary and clerk (he was also Mrs. Lyman's father).

Jim Stevens, then just 25 years old, was hired from the Goudy & Stevens Yard in East Boothbay in early 1941 to lead the process of lofting of vessels for the new company.

Hendel began working at Camden Shipbuilding on November 22, 1941, earning $99 per week. His Camden Shipbuilding office was in a small building at Eaton's Point, and there he took charge of the hull building department. During the war years, the yard produced 165' salvage rescue tugs (ATRs) and 103' APc personnel transports for the U.S. Navy. Also constructed were four 192' self-propelled, wood-hulled "Pine Tree" coal barges (MCc class), which the yard developed for the Navy using historic plans and models available there. Camden Shipbuilding prepared master drawings for the barges, and the drawings were then sent to other yards where about 300 other barges were built before the end of the war. The barges carried coal to New England ports for residential and industrial heating during the war years.[88]

Hendel received a deferral from active military service because of his work as a naval architect involved in the construction of Navy vessels.

In his audio interview, Hendel said he did not do any actual design work for the coal barges, but his job at the busy yard included buying the huge supply of materials, especially wood, needed for constructing the vessels built there. The yard employed more than 1700 men and women during the war years.

Hendel continued as an employee of Camden Shipbuilding until the end of the war in 1945, when he established his own naval architecture practice. He initially rented space at the Camden Shipbuilding yard, but after two fires in his office, he moved it to 30 Bay View Street in Camden.

Naval architect Geerd Hendel (center) and Camden Shipbuilding Co. owner William Petersen (right), along with John "Jay" S. Hanna (left) of Rockport, a long-time professional model-maker. Hanna is holding his model of the 40' ketch Freya. *Hanna's models are held in the nautical collections of the Smithsonian Museum, Mariner's Museum, Maine Maritime Museum, and among many other public and private collections.*
Barbara Dyer Collection, Walsh History Center, Camden Public Library.

After the war, Camden Shipbuilding refocused its business on building yachts and small pleasure craft. In 1946, former naval officer William "Pete" Peterson Jr. of Damariscotta was hired to run the yard and in 1948 purchased the "Bean Yard" property and managed it as the Camden Shipbuilding Co.

Hendel had many design commissions after the war and employed as many as four design assistants at some points. Among those he mentioned in his 1988 audio interview were Robert Jones, Dick Thomas, Günther Kleeberg, and Laura Pomeroy. George I. Dodge Jr., who had studied architectural drawing at the University of Maine, worked as a draftsman under Hendel's supervision from about 1944 to 1947.

Circa 1945: 18' aluminum centerboarders under construction at Camden Shipbuilding. The yard's construction of aluminum-alloy boats resulted from the influence of Hendel's experience with the metal.
Barbara Dyer Collection, Walsh History Center, Camden Public Library.

18' centerboarder sailing in Camden Harbor, probably one with an aluminum hull.
Geerd Hendel Collection, courtesy of his stepdaughter, Brigitte M. Gefrich.

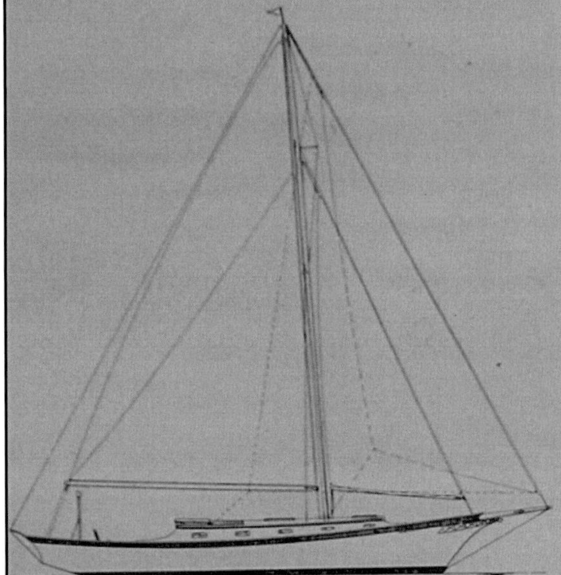

Sloop by Hendel

DESIGNED by Geerd N. Hendel of Camden, Me., this 33 foot auxiliary sloop is to be constructed next year for William E. Peterson, Jr., manager of the Camden Shipbuilding Co., Camden, Me. Her principal dimensions are: 33 feet overall, 25 feet on the water, 10 feet 6 inches beam, 5 feet 6 inches draft. Her total working sail area is 566 square feet.

This boat will have a fairly large displacement and is to be of solid construction. She is designed for use chiefly along the Maine coast for comfortable cruising and not racing. The arrangement is more or less standard, with double stateroom forward, then toilet and lockers followed by the main cabin and galley. Headroom in the main cabin is 6 feet 2 inches. The accompanying sketch shows the upper berth on the port side ready for night use, while on the starboard side the berth is folded out of the way.

The cockpit is roomy and comfortable, with a bulwark of not less than 5 inches in height around it. There will be no coamings to trip over, and the boat should have enough stability so that the rail will not be under water often.

The designer chose the raised deck type boat for the purpose of gaining space in the main cabin as well as deck space for stowing and working sails, etc.

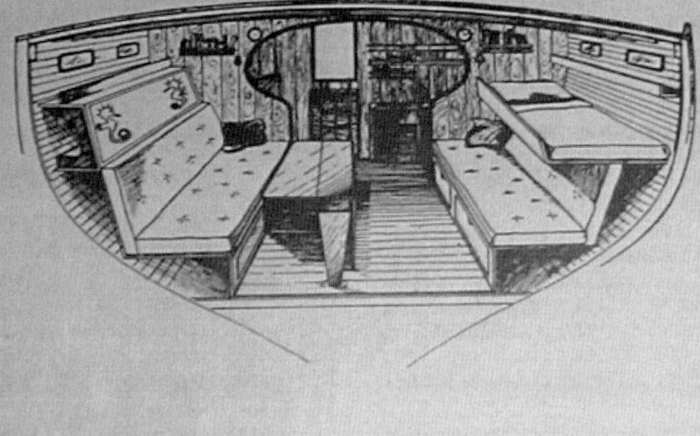

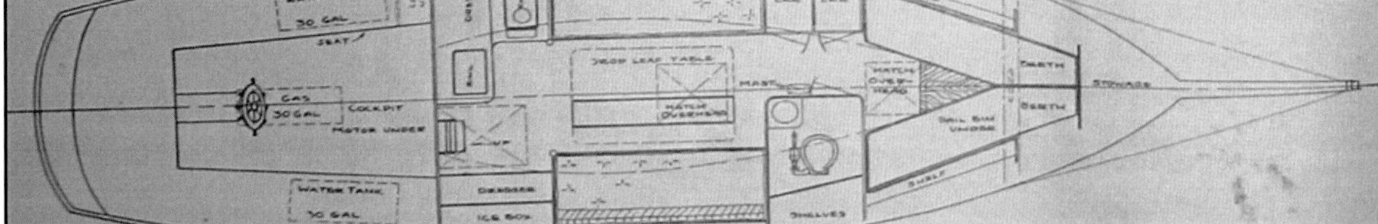

48

1946: a 33' sloop design for William "Pete" Peterson, owner of Camden Shipbuilding.

The Rudder magazine, December 1946.

1945: The 38′ Betselma *was built as a shrimp dragger for Ralph Wooster by Camden Shipbuilding. In the 1950s, she became a tour boat owned by Captain Les Bex, operating out of Camden Harbor. The seriously deteriorated vessel was taken to Artisan Boatworks in Rockport, Maine, in 2018 for planned restoration work, although her future remained in doubt.*
TakeMe2 Communications Collection.

1947: *Madelyn E*, the Rockland *Courier-Gazette* reported: "Fresh from her Camden Shipbuilding launching, this sleek Geerd Hendel–designed lobster boat and her Chrysler Crown gas engine get trialed before departure. This is yachting territory, so she'll fit better without the schooner-yacht background and after her iconic to-Matinicus canvas spray hood is in place."

1947: Madelyn E, *a 36′ lobster boat (hull F-4) built for Bradbury Young.*
Courier-Gazette Collection C177, Penobscot Marine Museum.

1947: A 55′ cruiser *Northern Cross* (Later *Romar-J* and *Penobscot*), Camden Shipbuilding hull Y-5. Built for Cary W. Bok, vice president of Camden Shipbuilding during WWII, *Northern Cross* was a cruiser of 55′ LOA designed by Geerd Hendel; she was launched on June 27, 1947.[89]

Northern Cross *under construction at Camden Shipbuilding.*

Geerd Hendel Collection, courtesy of his stepdaughter, Brigitte M. Gefrich.

Northern Cross.

Geerd Hendel Collection, courtesy of his stepdaughter, Brigitte M. Gefrich.

1947: 42' sloop *Gone Away*, Camden Shipbuilding & Marine Railway Co. hull (Y-6), was a Geerd Hendel design for Morris Cheston of Philadelphia. She was 30'6" on the waterline with a beam of 10'4". Keel and frames were white oak with African mahogany planking, and power was by a 46-hp Gray 162 Luger engine. Her mainmast was elliptical aluminum alloy made by Zephyr Products in New Kensington, Pennsylvania, the rigging was stainless steel, and she carried 845 square feet of sail. Accommodations were for four (see Chapter 8).[90]

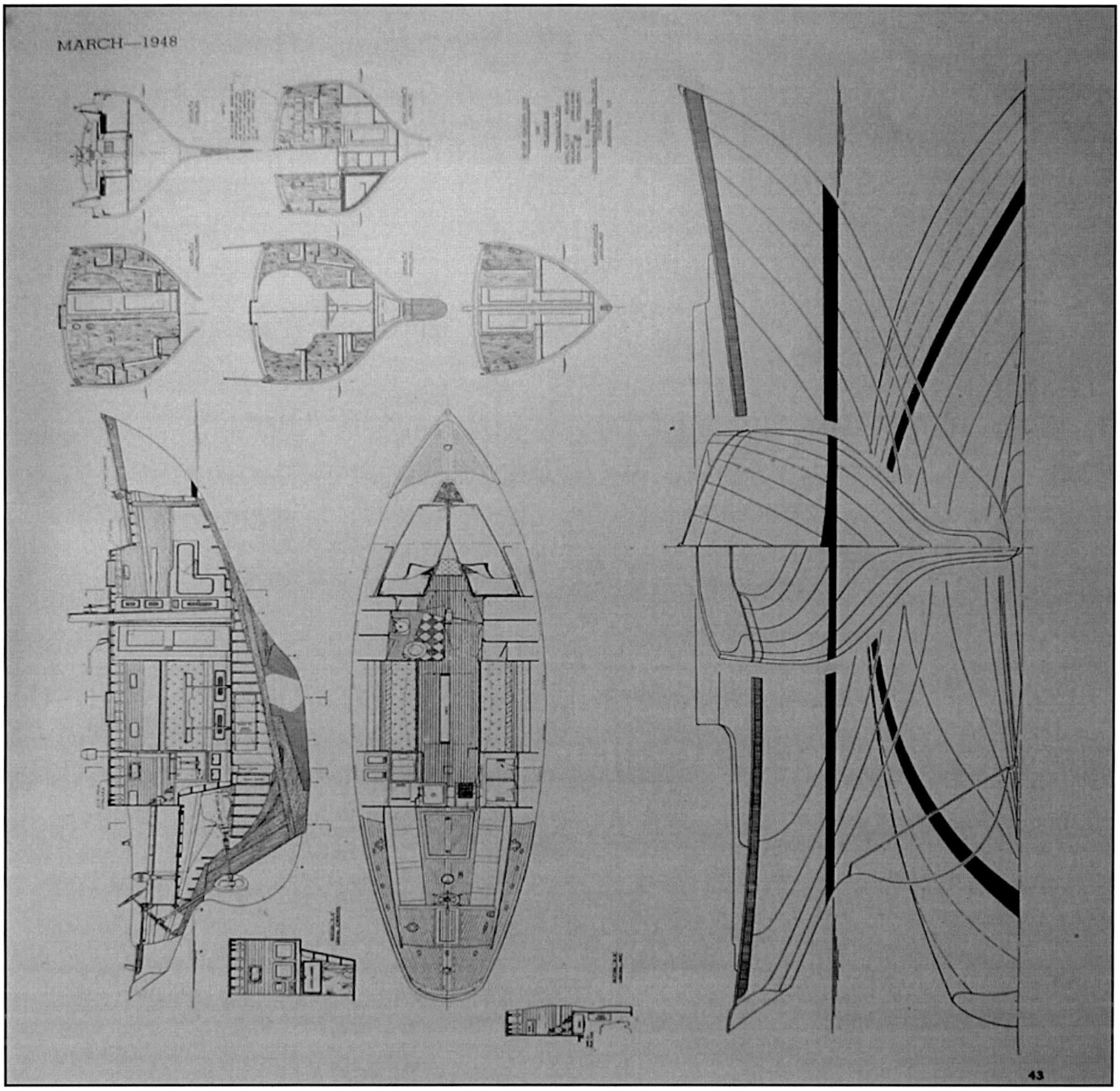

1948: 45′ sloop Alphard, *Camden Shipbuilding Company, Inc.*
Barbara Dyer Collection, *The Rudder* magazine, March 1948, Walsh History Center, Camden Public Library.

Alphard, crewed by the owner, Judge Curtis Bok, and his son Ben Bok of California; Dr. Morton A. Kravitz of Philadelphia and Norris Hoyt of Newport as mates; and Frank Crowley of Boston as cook, embarked on a six-week voyage to Plymouth, England. *Alphard* was re-rigged in 1953 as a yawl: eight feet were cut from her mainmast and a mizzenmast was added. The changes were to make her easier to handle for a crew of five on a planned transatlantic crossing to Spain, where rough seas and windy weather were anticipated. In fact, the vessel was dismasted in rough seas, and for several days it was feared that all hands were lost. A cable message, however, provided assurance that all were safe.[91]

Alphard *in 1954.*

Barbara Dyer Collection, Walsh History Center, Camden Public Library.

1949: 55' motor yacht Fantasy *(later* Murlyn II*) built for Donald Parson by the Camden Shipbuilding Company, Inc. (hull Y-9).*

Barbara Dyer Collection, Walsh History Center, Camden, Maine.

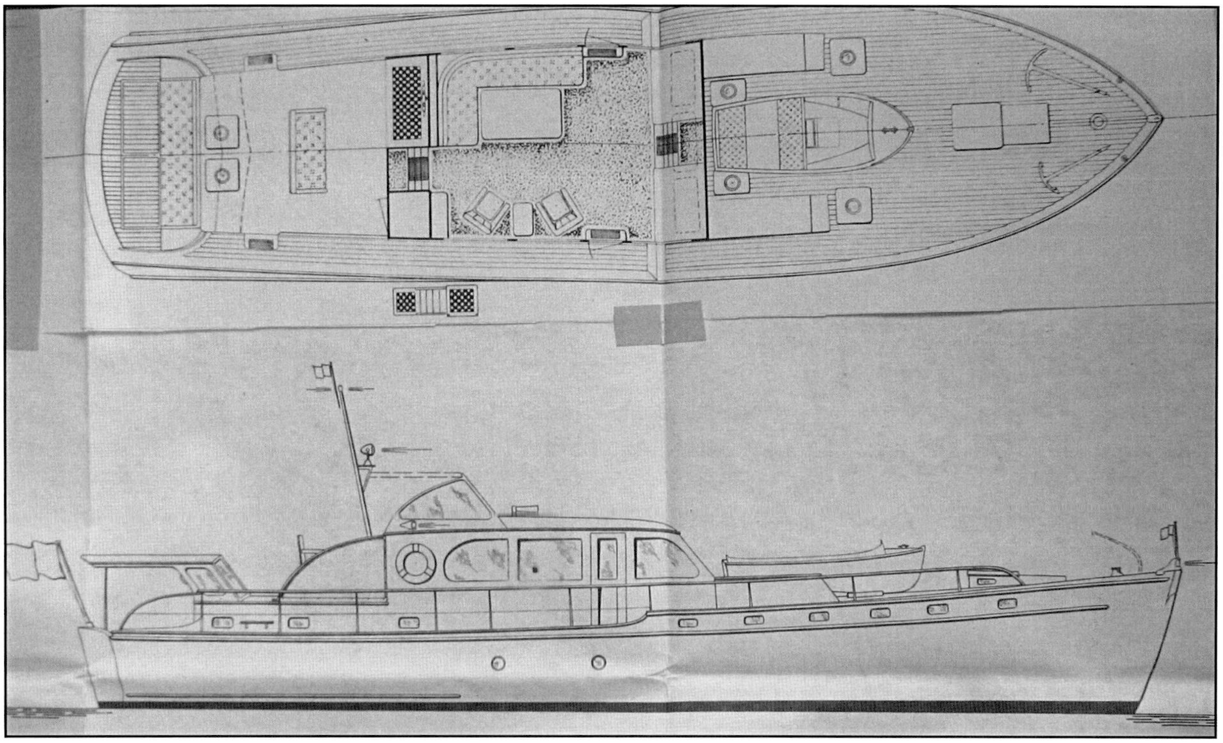

Fantasy *interior layout and design profile.*

Hendel Photograph Collection, Box MS-310, Maine Maritime Museum.

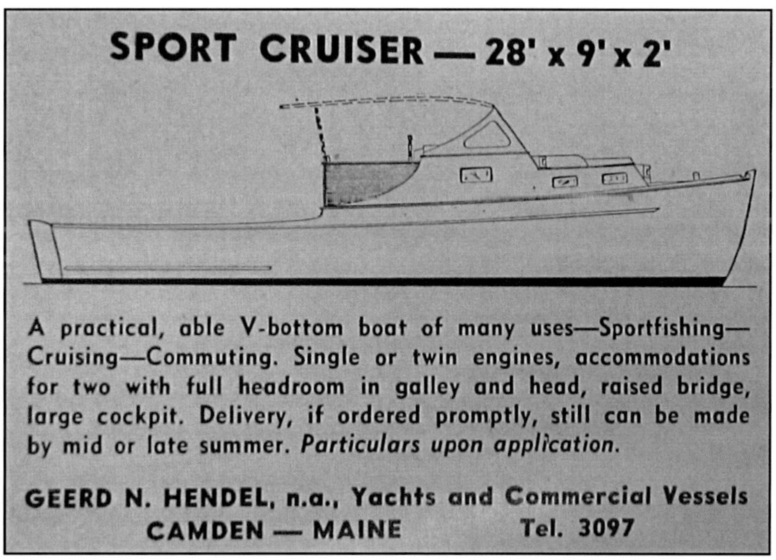

1949 sport cruiser.

Hendel Collection, Box 1, List #6, PL-19 98.106.4, Maine Maritime Museum.

A boat of this design was built for B. Devereaux Barker Jr. by Camden Shipbuilding in 1949.

1951: Sunbeam, *a 64 twin-screw diesel passenger boat; lines drawn by Hendel for the Deer Island Granite Corporation in Stonington, Maine.*
Hendel Photograph Collection, MS-310, Box 3, Folder 4, Maine Maritime Museum.

Anihita IV was a 70' yacht for Alfred Epstein, with a home port of Grosse Pointe, Michigan. Her 15 keel was oak, and she was double-planked with an inner layer of cedar and an outer layer of mahogany. Powered by twin 800-hp GM diesels, her top speed was 17 knots. The decks were teak, and the interior was beautifully appointed with mahogany-trimmed black walnut and mahogany furniture. A 110-volt electrical system was designed for the vessel.[92]

A 42' sport cruiser was designed by Hendel in 1947 for J.C. Aspinwall, but there are no records documenting her construction. Her lines are very close to those of *Wild Goose.*

1951: 70 diesel yacht Anihita IV, *Camden Shipbuilding Company, Inc. (hull Y-11).*
Geerd Hendel Collection, courtesy of his stepdaughter, Brigitte M. Gefrich.

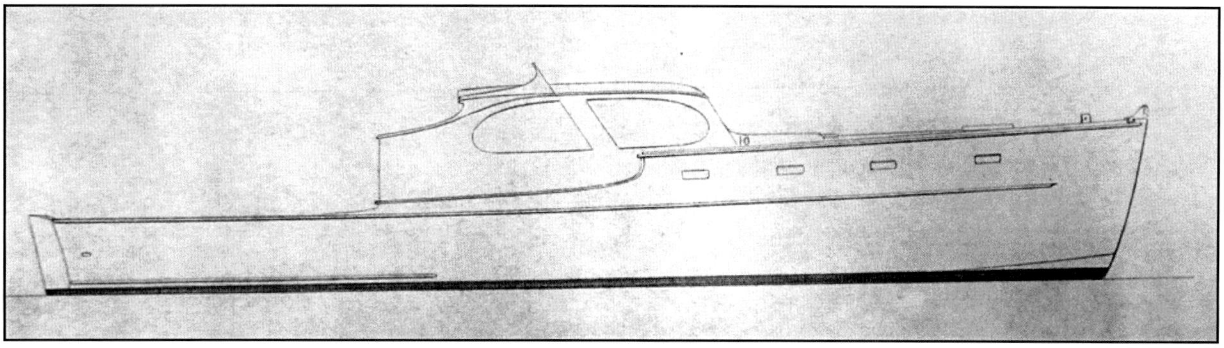

1947: 42' sport cruiser designed by Hendel in 1947 for J.C. Aspinwall.

Hendel Collection, Box 1, List #6, PL-19 98.106.4, Maine Maritime Museum.

Wild Goose.

Barbara Dyer Collection, Walsh History Center, Camden Public Library.

The 42' sport fisherman *Wild Goose* was built by Camden Shipbuilding (hull Y-13) in 1951 for W.H. Peters of Hackensack, New Jersey. She was sold to Ferdinand Eberstadt, however, while she was still under construction. She had quarters for two forward and a double stateroom aft.

She was designed to work in the Gulf Stream off the Florida Keys and in the West Indies, and four fishing chairs were installed in the cockpit. Her keel and frames were oak, and the hull was planked with mahogany over cedar. Powered by two 275-hp Hall-Scott Invaders, her estimated speed was 26 knots.[93]

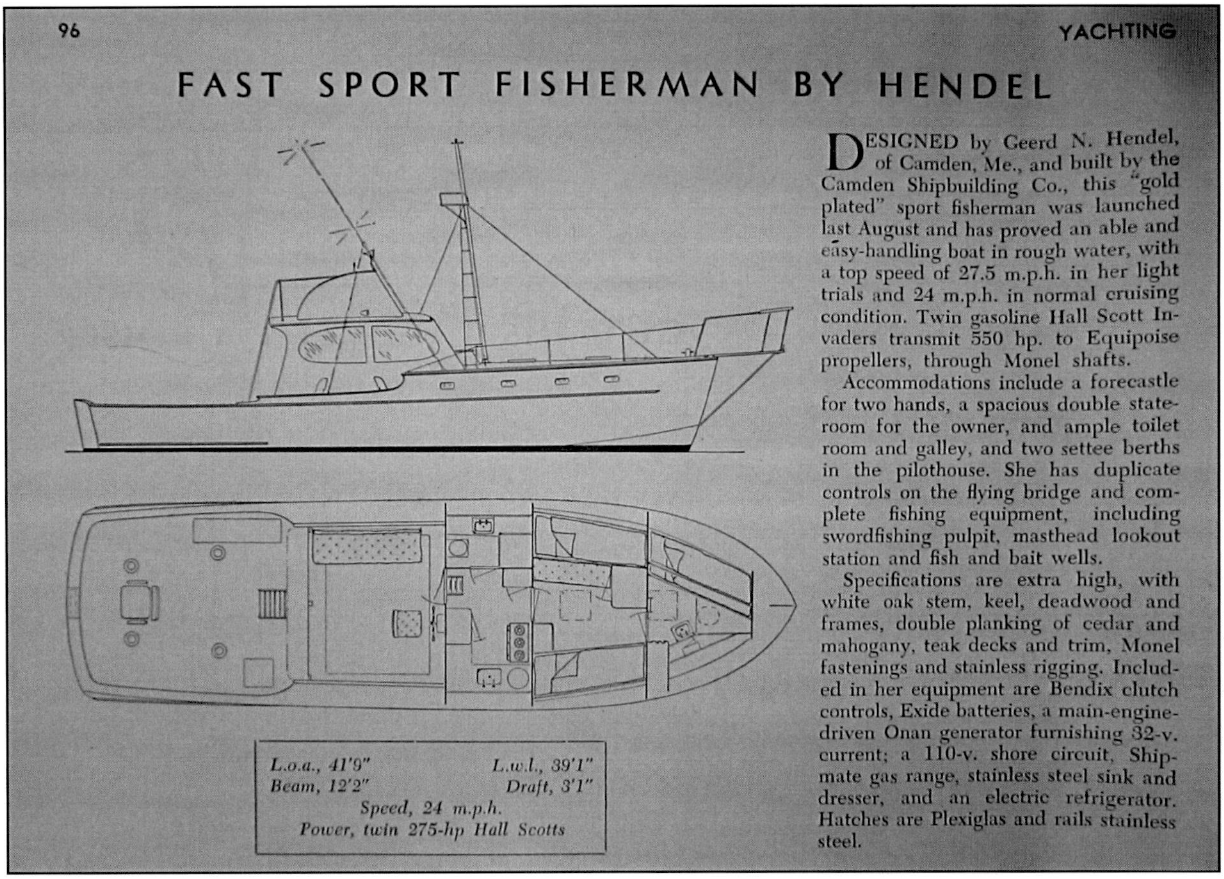

96 YACHTING

FAST SPORT FISHERMAN BY HENDEL

DESIGNED by Geerd N. Hendel, of Camden, Me., and built by the Camden Shipbuilding Co., this "gold plated" sport fisherman was launched last August and has proved an able and easy-handling boat in rough water, with a top speed of 27.5 m.p.h. in her light trials and 24 m.p.h. in normal cruising condition. Twin gasoline Hall Scott Invaders transmit 550 hp. to Equipoise propellers, through Monel shafts.

Accommodations include a forecastle for two hands, a spacious double stateroom for the owner, and ample toilet room and galley, and two settee berths in the pilothouse. She has duplicate controls on the flying bridge and complete fishing equipment, including swordfishing pulpit, masthead lookout station and fish and bait wells.

Specifications are extra high, with white oak stem, keel, deadwood and frames, double planking of cedar and mahogany, teak decks and trim, Monel fastenings and stainless rigging. Included in her equipment are Bendix clutch controls, Exide batteries, a main-engine-driven Onan generator furnishing 32-v. current; a 110-v. shore circuit, Shipmate gas range, stainless steel sink and dresser, and an electric refrigerator. Hatches are Plexiglas and rails stainless steel.

L.o.a., 41'9" *L.w.l., 39'1"*
Beam, 12'2" *Draft, 3'1"*
Speed, 24 m.p.h.
Power, twin 275-hp Hall Scotts

Wild Goose *profile and interior layout plan.*

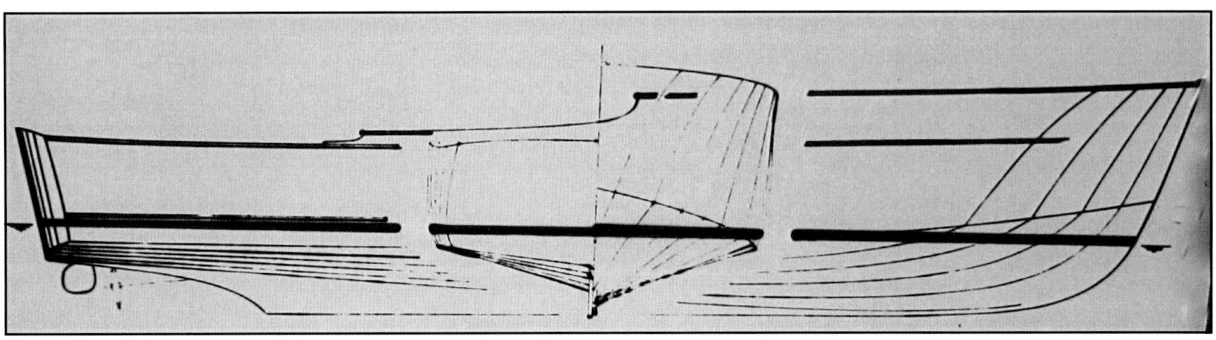

Wild Goose *hull profile.*
Yachting Magazine, January 1952.

22 FOOTER OF ALUMINUM

Latest in a series of aluminum-hulled pleasure boats and work craft from the boards of Geerd N. Hendel is a 22-foot all aluminum utility boat.

Fabricated entirely of aluminum at the Camden Shipbuilding & Maine Railway Company, Camden, Maine, the boat has attained trial speeds with her Chrysler Crown engine of better than 20 knots without pounding. At reduced speeds, the boat has been designed to make a complete 360-degree turn using the bow as a pivitol point! Her designer and builder claim that she'll ride level at all speeds. A flared bow and spray deflectors keep occupants dry in relatively rough seas, even at full throttle.

Construction is of riveted and welded design, ensuring maximum strength with a minimum of weight, according to marine engineers of Aluminum Company of America who participated in her trials.

Hendel's ongoing interest in building boats of aluminum alloy was evident in this 1948 news item from MotorBoating *magazine. These 22-footers were Hendel's "Bass Boat" design, with hulls of aluminum or wood and flexible cabin/cockpit layouts. The 1949–1952 bass fisherman production boat was available in three configurations from the Camden Shipbuilding Company.*

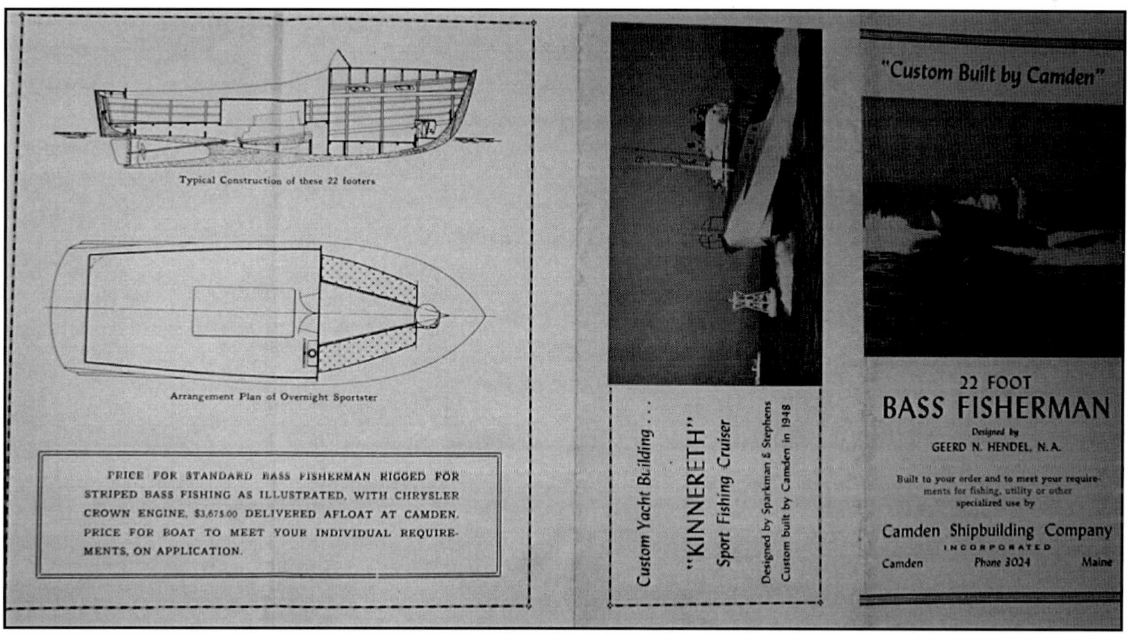

22' bass boat brochure showing configuration options for (1) standard bass fisherman, (2) utility fisherman with shelter, and (3) overnight sportster. Brochure text reads: "Keel, backbone, and frames, selected oak; planking, white cedar; oak seam battens; Everdur screw fastened; engine to suit; shaft, strut, and rudder fittings, Tobin bronze and Everdur. Standard equipment furnished, lights, horn, fire extinguisher, anchor, anchor rode, docking lines, fenders, four life jackets, registration numbers. Not mass produced—custom built to your order only. These boats are built to the same high standards and by the same craftsmen as our high grade yachts. Highly maneuverable at slow speeds. Will turn 360° at idle throttle without appreciable advance. Highly maneuverable at high speeds. Excellent performance in surf and around rocks. Level riding, no pounding, no rooting. Will maintain high speed in rough water safely and securely."

Hendel Photograph Collection, Box MS-310, Maine Maritime Museum.

The 22' bass boat *Chausser* (hull Y-10) was built by Camden Shipbuilding in 1949 for Mr. Gardiner. "Two more of these bass boats were built in 1952: the *Night Owler* (hull F-11) for Mr. Donald Owler and *Sea Coot* (hull F-12) for Mr. Winslow Hall."[94]

Edward A. Dodge, son of Geerd Hendel's 1944–47 drafting assistant George I. Dodge, recalled his father commenting that these bass boats were made unsinkable with Styrofoam flotation, which was unheard-of by the yard workers before that time.

Per Hendel's correspondence with ALCOA in 1954 regarding its request to have constructed a 22' launch based on his bass boat hull design, Hendel responded that Camden Shipbuilding had ceased

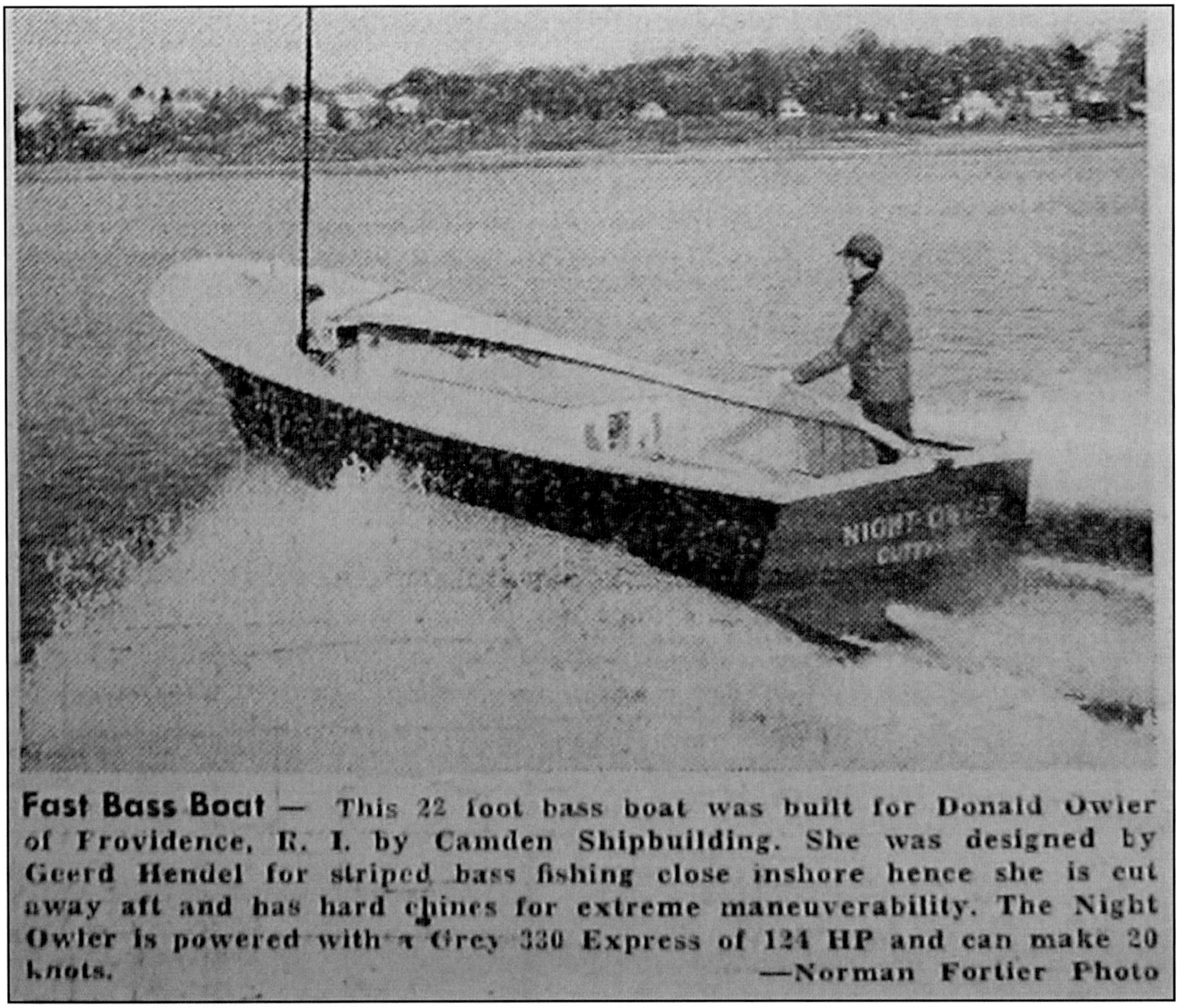

Fast Bass Boat — This 22 foot bass boat was built for Donald Owler of Providence, R. I. by Camden Shipbuilding. She was designed by Geerd Hendel for striped bass fishing close inshore hence she is cut away aft and has hard chines for extreme maneuverability. The Night Owler is powered with a Grey 330 Express of 124 HP and can make 20 knots. —Norman Fortier Photo

22' bass boat Night Owler *built in 1952 by Camden Shipbuilding.*

Hendel Collection, *Maine Coast Fisherman*, February 1953, Maine Maritime Museum.

building aluminum boats about 1949 because riveting aluminum hulls was expensive and time-consuming. The General Foods Corp. yard in Rockland was at that time welding aluminum hulls using newly developed technology, and the launch was constructed there.

Built for Allison Layton, an experienced sword fisherman, *Porpoise* had oak keel and framing, double-planked mahogany over cedar, teak decks, and accommodations for six.[95]

The 22' bass boat Russell B *built of aluminum for Earl Boardman in 1952 by Camden Shipbuilding.*
Geerd Hendel Collection, courtesy of his stepdaughter, Brigitte M. Gefrich.

Joel White (1930–1997) of Brooklin, Maine, a naval architect and boatbuilder who was considered one of the country's foremost designers of wooden boats, wrote glowingly of *Porpoise*'s design:

> *Porpoise's* hull is nearly perfect; I can think of no change that would make her better. Her superstructure is nearly so. Geerd Hendel, who designed this paragon, managed to blend the two elements, hull and superstructure, into a unified whole that delights the eye, or at least my eye....But it is not her bowsprit or mast that makes her design so unique or distinctive—remove them and she would look better, if anything. The long, continuous flow of the sheerline, starting from a relatively low freeboard at the after end of the superstructure and rising steadily to the flaring clipper bow profile, is one element of the design that gladdens the eye. The low freeboard is made to appear even lower by the continuous guardrail at the deck edge.

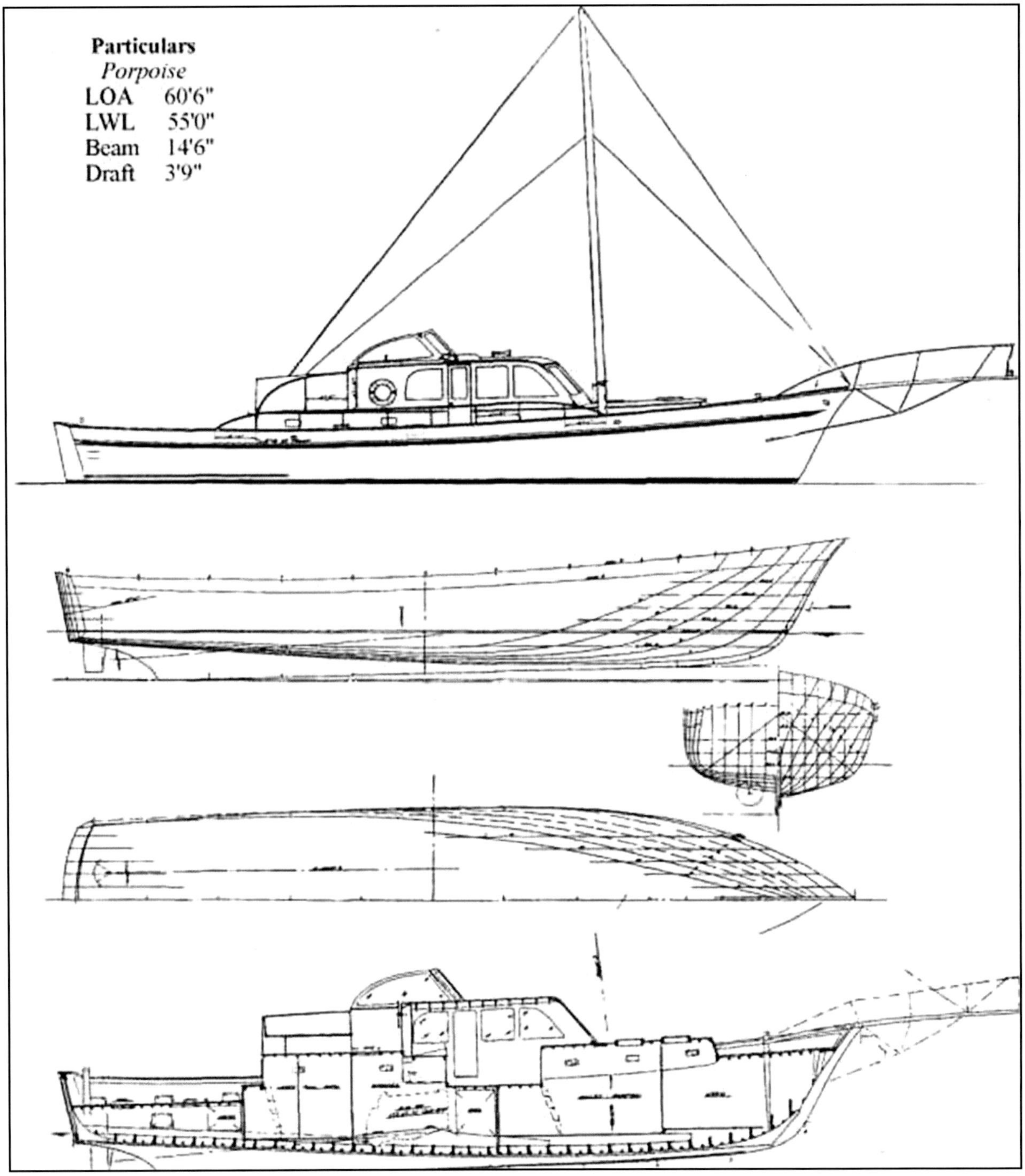

1952: 61' sport fisherman Porpoise *(later* Gansett II*), Camden Shipbuilding hull F-10.*

Barbara Dyer Collection, Walsh History Center, Camden Public Library.

> The other element that contributes so much to the overall impression of the boat is the low, short deckhouse structure. This superstructure is very much concentrated amidship[s], leaving a good deal of hull showing alone at the bow and the stern. The midships handrail, with a varnished teak rail cap set on bronze stanchions, and a canvas weather cloth laced between the handrail and the cap rail, was designed to give one a feeling of security while walking the side decks. It also does much to lower the apparent height of the deckhouse. I realize that I have spent a lot of words analyzing the appearance of this boat, but beauty is so rare and nowadays so neglected in the design process that it seems worthwhile to consider.[96]

White's review of the vessel also contains greatly expanded comments concerning the details of the vessel's construction.

Porpoise.
Geerd Hendel Collection, courtesy of his stepdaughter, Brigitte M. Gefrich.

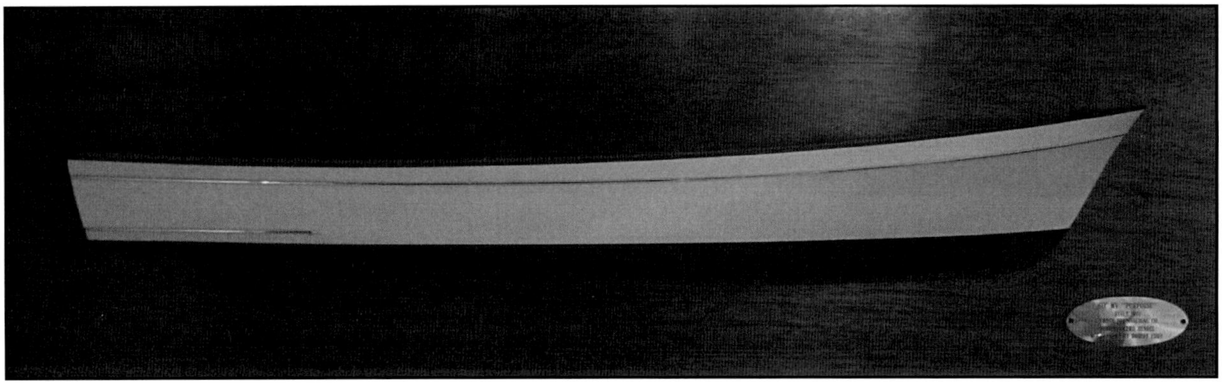

Porpoise *half-hull model by Rob Eddy.*

Geerd Hendel Collection, courtesy of his stepdaughter, Brigitte M. Gefrich.

Porpoise *graced the cover of the November 1952 issue of* The Rudder *magazine.*

Barbara Dyer Collection, Walsh History Center, Camden Public Library.

68' diesel yacht Sooner *(later* Voyageur, Lady Grace, *and* Lady Ferrell*) built in 1954 by Camden Shipbuilding (hull Y-15).*
Barbara Dyer Collection, Walsh History Center, Camden Public Library.

Sooner was built for Julian Davis and was heavily constructed for offshore cruising, with a hull built over 3" x 3" frames, 10" on center, and planked with 1½" mahogany. Power was two GM 400-hp diesels, electronics were extensive, and she could carry 1600 gallons of fresh water and 3000 gallons of fuel.[97]

According to Barbara Dyer, her elegant appointments included white silk curtains and red carpeting.

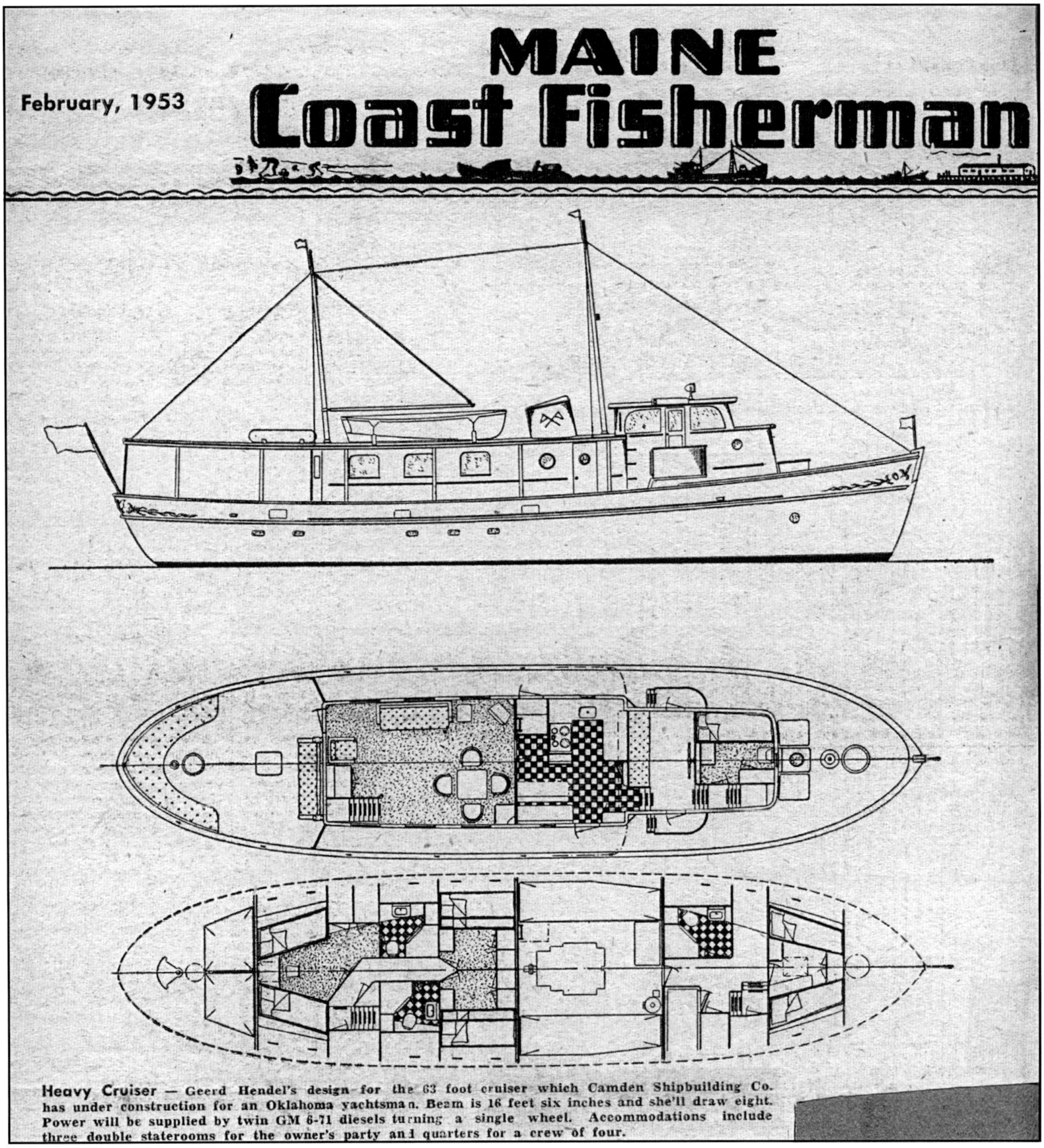

February, 1953

MAINE Coast Fisherman

Heavy Cruiser — Geerd Hendel's design for the 63 foot cruiser which Camden Shipbuilding Co. has under construction for an Oklahoma yachtsman. Beam is 16 feet six inches and she'll draw eight. Power will be supplied by twin GM 6-71 diesels turning a single wheel. Accommodations include three double staterooms for the owner's party and quarters for a crew of four.

Sooner *profile and interior layout. The text reads as follows: "***Heavy Cruiser***—Geerd Hendel's design for the 63-foot [68' is correct LOA] cruiser which Camden Shipbuilding Co. has under construction for an Oklahoma yachtsman. Beam is 16 feet six inches and she'll draw eight feet. Power will be supplied by twin GM 6-71 diesels turning a single wheel. Accommodations include three double staterooms for the owner's party and quarters for a crew of four."*

Barbara Dyer Collection, Walsh History Center, Camden Public Library.

The 73' *Admiral Blake* was launched at Camden Shipbuilding Co. in the spring of 1956 for John Blake, a New York television and radio executive. Its beam was 18' and power was by twin GM diesels. The hull was mahogany planked, and her decks were Alaskan cedar. She took just over a year for the yard's 43 employees to build. She was designed to carry two tenders: a 15' Old Town lapstrake outboard and an Old Town skiff. The red, white, and blue vessel was named for John Blake's seventeenth-century ancestor, British Admiral Robert Blake, who twice defeated the Dutch fleet and also commanded the first successful attack of seashore castle fortifications.

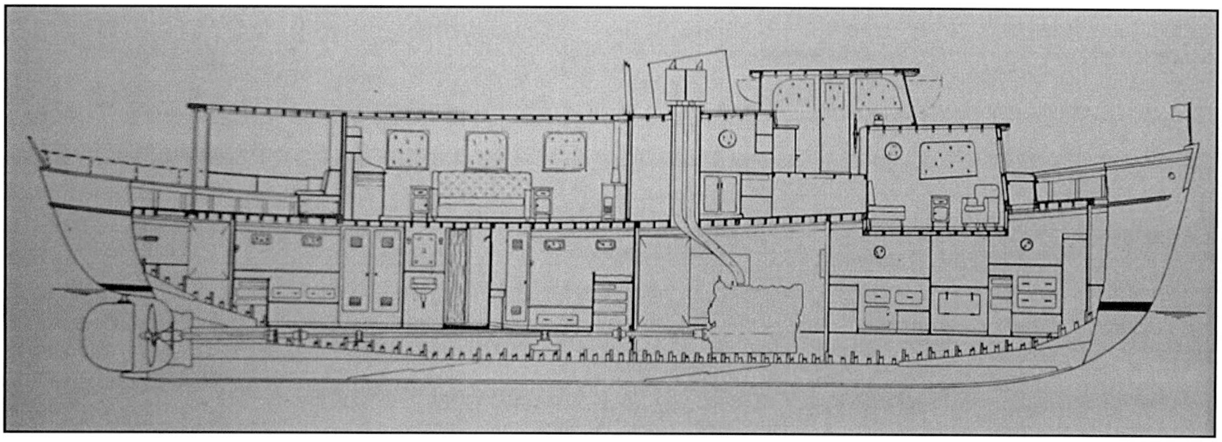

Admiral Blake *cut-away plan.*
Geerd Hendel Collection, courtesy of his stepdaughter, Brigitte M. Gefrich.

Admiral Blake *under way.*
Geerd Hendel Collection, courtesy of his stepdaughter, Brigitte M. Gefrich.

Admiral Blake's *stern deck.*
Geerd Hendel Collection, courtesy of his stepdaughter, Brigitte M. Gefrich.

1956: 73' yacht *Admiral Blake* (later *Dragoon, Duke and Duchess, Varamista, The Ritz, Las Tango,* and *Arctic Fjord II*) built by Camden Shipbuilding (hull Y-16).

Hendel Photograph Collection, MS-310,
Box 1, Folder 4, Maine Maritime Museum.

Pearl Necklace *renamed* String of Pearls.

Website of Golden Isles, Georgia, Convention and Visitors Bureau.

1957: 71' cruiser *Pearl Necklace* (later *String Of Pearls* and *Cloister Belle*), Camden Shipbuilding Company (hull V-1).

Built for the Marine Development Co. of Providence, Rhode Island, this handsome 71' boat's hull was planked with $\frac{7}{8}$" mahogany over $\frac{3}{8}$" cedar, over $1\frac{1}{4}$" x $1\frac{3}{8}$" oak frames. She was fitted with full electronics and carried two riding/steading sails. Her accommodations were designed for four and a crew of three.[98]

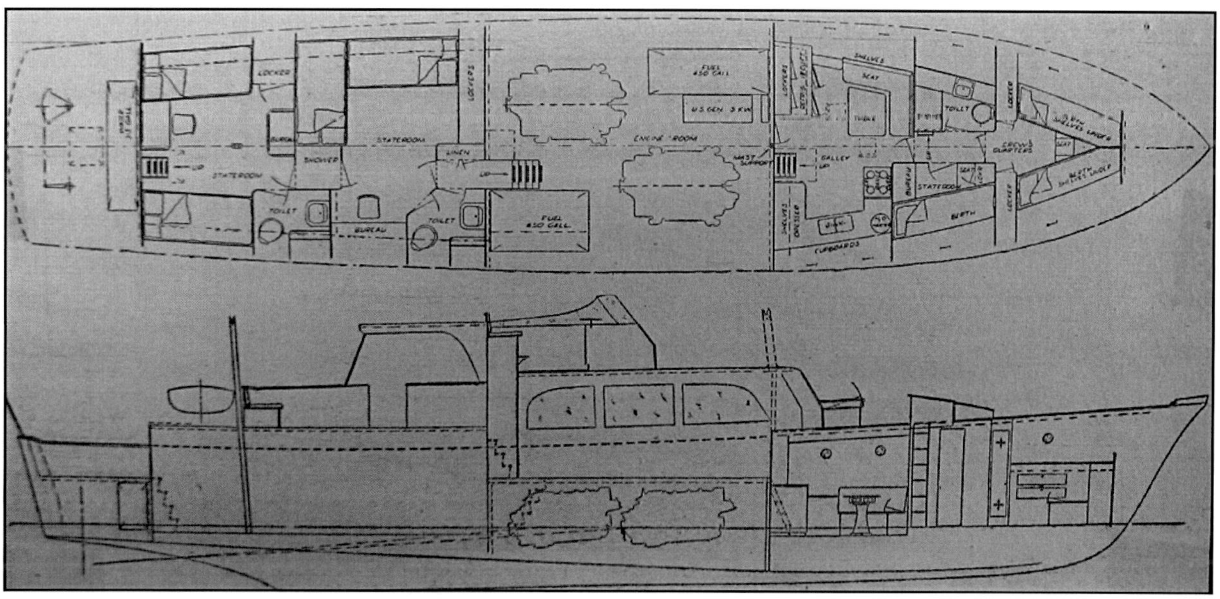

Pearl Necklace *interior layout and profile.*

Barbara Dyer Collection, Walsh History Center, Camden Public Library.

A description of the vessel in the *Portland Press Herald* says that she was designed for cruising and fishing from the waters of New England to the Caribbean. Hendel tested her hull at the Stevens

The Pearl Necklace *has had several owners over the years. Two of the owners altered Hendel's flying bridge design, with some of the alterations unfortunately detracting from her original lines.*

Geerd Hendel Collection, courtesy of his stepdaughter, Brigitte Gefrich.

Pearl Necklace, *renamed* Cloister Belle, *in a 2013 charter ad.*

Register of Wooden Boats, *WoodenBoat* magazine.

Institute of Technology in Hoboken, New Jersey, and it was rated at 22 knots with 1000 hp, and cruising speed was estimated to be 20 knots with its 947 hp from dual GM 6-71 engines driving twin screws. The port engine was placed in a rear position to bring it close to the center of the craft. Fuel tanks were also staggered to balance weight differences. The vessel had safety glass all around. To save weight and at the same time gain strength, the C decks were faced with ¼" teak over ½" fir plywood, which was bolted through aluminum deck straps set in rabbeted king planks. The *Pearl Necklace* had ketch-rigged steadying sails. Its small boat equipment included a 15' aluminum fishing dinghy, a 10' aluminum pram, and an 8' Plexiglas pram for underwater viewing. The vessel had a large salon and a 19-cubic-foot galley freezer. Electricity was from a U.S. Motors generator set of 32 volts DC, supplying 5 kW. All steering was done through the flying bridge, with hydraulic steering from Crowell Designs and a Sperry automatic pilot. The rudders and screws were fabricated by the Columbian Bronze Corp., Freeport, Long Island. Exquisite wood finishing included butternut in the owner's quarters and rift ash in the salon and guest stateroom.[99]

As *String of Pearls*, the classic yacht continued in active use. Captain Brian Quin bought the yacht in 1985, restored it in Newport, Rhode Island, and then chartered her throughout New England for 19 years. From 2004 to 2018, as *Cloister Belle*, she served resorts on the Georgia coast as an available charter.

The website www.yachtcharterguide describes *Pearl Necklace* thus:

> Built in 1957 as an original Geerd Hendel design, *Pearl Necklace* was a husband's gift to his wife, who had requested a simple strand of pearls for their anniversary. The yacht was built in Maine at Camden Shipyards [sic], one of America's premier yards at the time. Painstakingly crafted of double-planked mahogany on cedar and powered by four diesels, she was one of the fastest privately owned yachts of her size, making better than 24 knots.... Historically significant, she has hosted business, political, and socialite figures and made a variety of TV and film appearances.
>
> Fully restored and those exciting days but a memory, *Pearl Necklace* is now offered for charter as one of New England's best-loved classic motoryachts.
>
> USCG Certified, she carries up to 49 guests in stabilized comfort with uniformed first-class service, providing a unique and elegant setting that captures the splendor of a bygone era. A professional, congenial crew of 3 promises you a memorable day on the water.
>
> Its business is upscale private parties, corporate functions, dinners, weddings, ash scatterings, spectator/sponsor yacht for sailboat races, etc. The boat has been chartered for movies, commercials, TV shows, and photo shoots... *Pearl Necklace* is chartered by the hour.
>
> ...[T]eak decks, brass fittings, varnished trim and traditional appointments highlight the distinguished features of this lovely classic...

Shada *shown with owner Daniel R. Sortwell Jr. and his eldest daughter, Betty Ross, aboard.*
Hendel Photograph Collection, MS-310, Maine Maritime Museum.

1958: *Shada* (pronounced "shadda") was a 33' twin-screw cruiser built by Camden Shipbuilding. She was owned by Daniel R. Sortwell Jr., who lived in Wiscasset in what is now the Nickels-Sortwell House Museum at 121 Main Street. He kept *Shada* on a mooring near the present Boothbay Harbor Yacht Club in Boothbay Harbor, where she was primarily used for family cruising and picnics.

Shada is a boat name used by three generations of the Sortwell family for their boats. Current Camden resident Edward "Terry" Sortwell; his grandfather, Daniel R. Sortwell Jr.; and his great-grandfather, Daniel R. Sortwell Sr. According to the Sortwells, the word *shada* appears in *The Song of Hiawatha* by Henry Wadsworth Longfellow in reference to the pelican.

Camden Yard Building Fast, Agile Sportster

CAMDEN, ME. – Camden Shipbuilding Co. is building a new design by Geerd Hendel on which the firm has a competitive price in the sport fisherman class. She's an able-looking 33-footer – a "Sportster" able to get up and go, yet maneuverable enough so that she can be used by sportsmen along the coast.

A unique feature is a cut-away skeg which allows more water to get to the wheel, gives a shorter turning radius, and makes the spade-type rudder more effective. With a 225-horse Chrysler driving a single screw, the Sportster is expected to do better than 20 knots and cruise easily at 16.

The hull has more deadrise than is usual in most Maine-built boats. This has been done to give the boat an easier motion, make her faster, and make her bank better on turns.

The arrangement layout is flexible, and the boat can be rigged for sport fishing, as a day boat, or a week-end cruiser, whatever the owner desires. The present boat, which is being built "for spec", is set up to sleep two in the forward cabin, which contains galley and enclosed head. The galley has an alcohol stove and built-in ice box. Topsides, the boat features an open bridge with windshield and portable Navy top. She has a big cockpit useful for day cruising and fishing.

Hull construction will include oak keel and 1¼" x 1½" oak frames on 9" centers. Planking is 3/4" mahogany, and deck and cabin are of plywood covered with plastic. She will have a 30-foot waterline, 9-foot 6-inch beam, and draft of 2-feet 10-inches.

1958: 33' "Sportster" cruiser available at Camden Shipbuilding Company, Inc. A Sportster could be arranged to sleep four, and tanks provided capacity for 140 gallons of gasoline and 25 gallons of fresh water.

Barbara Dyer Collection, Walsh History Center, Camden Public Library.

1960: Sabrina, *a 33' Sportster built by Camden Shipbuilding (hull Y-17) for Paul Jones, MD, of Union, Maine. Some records show her name as* Sabina.

Barbara Dyer Collection, Walsh History Center, Camden Public Library.

Project.

Barbara Dyer Collection, Hamilton Hall photo, *The Camden Herald*, October 16, 1958, Walsh History Center, Camden Public Library.

1958: *Project* (later *Huntress*) was a variation of Hendel's 33' Sportster design described on the previous page. She was built at Camden Shipbuilding for Mr. and Mrs. William Brown of Camden. Brown sold *Project* to Duane Doolittle [founder of *DownEast* magazine] of Camden who owned her for many years. Here she's shown on the ways with launching assistance provided by *Susie*, the Camden Shipbuilding yard's workboat. The *Project*'s hull construction was ⅜" mahogany over 1¼" x 1½" oak frames 9" on center. Her interior layout was a stateroom for two, a main cabin with convertible lounges, living and dining space, and a galley. Hydraulic steering was installed on the port side of the cockpit, and a depth finder was included among her electronic equipment. Powered by a 275-hp Packard Clipper freshwater-cooled engine with a Walthers V-drive and a 1½:1 reduction, the boat was expected to make speeds in excess of 20 knots and have a cruising speed of 16 knots. Her prop was a 17" x 17" Columbian.[10]

A first for the Camden yard was the use of fiberglass to cover all her deck and deckhouse surfaces.

At some point *Project* was renamed *Huntress*. When it was offered for sale in the 1980s by yacht brokers Cannell, Payne, and Page, *Huntress* was described as having been rigged with a mast and boom in the mid-1970s for commercial trawling, and an air-cooled Lister diesel engine was installed. It was never used as a trawler, however, due to illness of its owner. Hendel, in a September 6, 1990 letter to the prospective buyers of *Huntress*, John and Susan Williams of North Haven and Annapolis, Maryland, shared his 1984 correspondence about the boat, which identified it as one of his Sportster designs. *Huntress* was at that time stored in a yard in Essex, Connecticut.

Huntress *(formerly* Project*).*
John and Susan Williams Collection.

Alster, *Hendel's 1970 personal Sportster with a customized cabin. (See Chapter 9 for more detail.)*
Geerd Hendel Collection, courtesy of his stepdaughter, Brigitte M. Gefrich.

Windless.
Barbara Dyer Collection, Walsh History Center, Camden Public Library.

1961: *Windless,* a 40' motor yacht by Camden Shipbuilding Company (hull Y-21). The flying-bridge cruiser was built for Anthony Winslow Jr. Mahogany-planked with teak decks, she had accommodations for four. Power was by two GM 200 diesels, and she had full electronic systems and navigational aids.[101]

1962: *The Frances,* a 40' auxiliary ketch built for Philip E. Merryman of Trumbull, Connecticut, was launched by Camden Shipbuilding on December 27, 1962 (see Chapter 8 for more detail). Although *The Frances* was her name at launching, she's often referred to simply as *Frances.*

CHAPTER ELEVEN

Tugs and Towboats

Hendel's experience in tugboat design began around 1932 when BIW had won a contract for a New York harbor tug for the Turecamo Towing Company. The 93' diesel tug *Turecamo Girls* was launched in May 1933 at BIW.

A similar tug, *Turecamo Boys*, was built in 1935–36 by Rice Brothers Shipbuilding when Hendel lived in Boothbay Harbor. Designed by naval architect Merritt Demerest of Jersey City, New Jersey, at 115' she was slightly larger than *Turecamo Girls* and was powered by a 650-hp diesel engine (www.tugboat-information.com).

In 1936, Hendel was employed in the drafting/design process for *Ranger* and resided in Boothbay Harbor. He mentions this tug being built at Rice Brothers in a letter dated November 9, 1935.

In 1949, the wooden-hulled 45' tug/towboat *Mirage*, designed by draftsman and naval architect Stanley Potter, was built at Camden Shipbuilding (hull WB-2) for Russell Spangee, and its home port was to be Norwalk, Connecticut. It's certain that Hendel was well aware of this vessel.

Turecamo Boys.

Hendel Photograph Collection, Captain Howard L. Wentworth Sr. photograph, Box MS-310, Maine Maritime Museum.

Bigtow and *Middletow*

Hendel entered into log-boom towboat design contracts with Central Maine Power Company (CMP) in the late 1940s. His familiarity with the Turecamo tug designs, *Mirage*, and his fishing trawler design work throughout the 1940s likely supported CMP's decision to award their logging towboat design work to him.

CMP ordered two log-boom towboats designed by Hendel that were built to work on Flagstaff Lake, located in Somerset County and Franklin County, Maine, where the North and South Branches of the Dead River join in the lake. The Long Falls Dam had been constructed in 1950, enlarging the natural Flagstaff Lake and turning it into a reservoir used for hydropower electricity production by regulating the flow of the Dead River into the Kennebec River. At the time, river log drives were still the primary means of delivering timber to the pulp mills downstream.

The dam is 45' high and 1339' long at its crest. River-based log drives ceased in the mid-1970s, and the dam's purpose now is to regulate and augment water flows that are used by eight downstream Kennebec River hydropower facilities and to control flooding; no electricity is generated there.

1950: *Bigtow* was a 40' steel tug owned by Central Maine Power Company and used as a logging towboat on Flagstaff Lake. It was built at the General Foods Corporation Yard in Rockland, Maine.

Bigtow's overall dimensions were 40.2' LOA x 12.7' beam x 4.6' draft; displacement was approximately 32,000 pounds. It was designed with a tunneled stern and sturdy guards to protect the propeller from pulp logs because it was meant to work close to shore or amid log rafts in shallow waters. Although built for CMP, it was operated by CMP's subsidiary, the Kennebec Log Driving Company. Designed by Hendel in December 1949, this vessel was destined to tow log rafts of 80,000 to 100,000 cords of pulpwood annually.[102]

Bigtow.

Hendel Photograph Collection, Box 1, Folder 12, Maine Maritime Museum.

Bigtow was powered by a 125-hp Caterpillar D-1300 marine diesel engine with a 2:1 reduction gear for the propeller. The engine was equipped with a front-end power takeoff geared through a three-speed transmission to operate a warping winch down in the engine room. The winch drum had a capacity of 4000' of 5/8" steel cable. Installed on the deck at the bow was a heavy roller with fairleads over which the warping cables ran to the winch below. A 650-pound warping anchor was to be shackled to the warping cable. In moving a log raft from the staging area out to deeper water, the *Bigtow* would drop anchor and play out the warping cable as she backed down and secured the logs. The warping winch would then pull in on the cable, moving the log raft out into deeper water. She was equipped with a Maxim muffler silencer, a bilge pump, and a fuel tank with a 330-gallon capacity. The portable 1500-watt generator was driven by a single-cylinder air-cooled engine and housed in the engine room.[103]

Rockland Historical Society, Rockland, Maine.

The 28' 1951 steel tug *Middletow* was also built for CMP at the same yard as *Bigtow,* and it was also powered by a Caterpillar marine diesel engine. *Middletow* was used on the Dead River in the Upper Kennebec River basin for rafting timber.

Middletow.
Hendel Photograph Collection, Box 1, Folder 35, Maine Maritime Museum.

"This tug was specially designed by naval architect Geerd Hendel, Camden, Maine, during August 1950, and followed the basic design of the *Bigtow* except it was 17' shorter....The tug worked in concert with the *Bigtow* in towing boomed logs and provided extra power when needed to haul exceptionally heavy rafts or when strong headwinds prevailed. Her shallower draft made it possible to work in shallow areas of the lake, gathering up logs and forming rafts where the *Bigtow* could not proceed safely."[104]

In 1952, the 24' *Kennebec No. 3* was built by the General Foods Corp. yard and launched for the Kennebec Log Driving Co. The design plans for this tug are in Hendel's plan archives at the Maine Maritime Museum.

Big Squaw, Little Squaw, and *Papoose*

CMP had three log-towing vessels built to handle log drives though Indian Pond. The *Big Squaw, Little Squaw,* and *Papoose* were also operated under the responsibility of the Kennebec Log Driving Company, which was chartered in 1837 to manage log drives on the Kennebec River.

In the years 1952–55, CMP constructed Harris Station, located on the upper Kennebec River at Indian Pond about 12 miles southwest of Moosehead Lake. Prior to Harris, there was a small wooden dam at the outlet of Indian Pond that served to regulate water flow for log drives and provide a crossing point on the Kennebec. Indian Pond originally covered about two square miles, but its size increased to 5½ square miles with the construction of Harris Station. The depth of the pond also increased by about 20', and the total impoundment is now nine miles long. Harris Station dam is the largest dam in Maine, standing 175' tall and spanning 270' across the Kennebec River gorge.

Harris Dam, which impounds the Kennebec River's Indian Pond just below Moosehead Lake's East Outlet.
Collections of Maine Historical Society.

When the dam was constructed, logging companies working in the area were still actively using the Kennebec River for log drives. To accommodate this activity, a large log sluice was built on the western end of the dam to allow logs to pass Harris and continue downriver toward the mills. In the 1950s, about 84,500 cords of wood passed through this sluiceway every year. There was also a large sluiceway about a half mile downstream near the confluence with Chase Stream. The last log drive in the continental United States occurred on the Kennebec in 1976, beginning in Moosehead Lake and continuing to Winslow, a distance of about 100 miles. With the passage of the federal Clean Water Act of 1972, and a state law based on environmental considerations, log drives ceased in 1976. Pulpwood had clogged lakes and rivers as the logs made their way downstream, which led to many water quality and recreational issues.

1954: The Big Squaw, *a steel log-boom towboat, was built at the General Foods Corp. yard in Rockland, Maine.*[105]
Rockland Historical Society, Rockland, Maine.

Papoose *and* Little Squaw.
Hendel Photograph Collection, Box 1, Folder 11, Maine Maritime Museum.

Big Squaw, 44'6" in length with a beam of 12' and drawing 5' of water, was powered by a 120-hp Caterpillar D-1300 diesel engine. She carried a 300-lb. Danforth anchor with a hand winch to provide "hold" while log booms were set up. A small galley held a Shipmate stove, and there were lockers for the crew's gear. There were berths for a crew of four. She was trucked to Indian Pond.

The 27' 1954 steel log-boom towboats *Little Squaw* and *Papoose* were also built for Central Maine Power Company at the General Foods

Corp. yard in Rockland for log drives on Indian Pond. Each was powered with a 38-hp D-311 Caterpillar diesel engine, had a length of 27', a beam of 9'8", and drew 3'8". An open boat with stick steering and no cabin, they were used to gather logs into booms in areas too shallow for the *Big Squaw* to operate.[106]

Nibroc

1953: The 43' log-boom tug *Nibroc* was built at the General Foods yard for the Brown Paper Company, Berlin, New Hampshire, and named for their paper towel product. The tug was powered by a 100-hp Buda diesel engine and was designed with crew bunk rooms and a galley. She was hauled by truck in two sections to Aziscoos Lake in northwestern Maine near the New Hampshire border and reassembled there. Hendel noted that the Brown Paper Co. required its towboats to be designed with narrow beams to facilitate trucking them from lake to lake.

Nibroc was featured on the cover of *The Brown Bulletin*, Vol. 2, Number 11, June 1954, p. 2, with this caption: "The new Brown Company tugboat *Nibroc* tows a boom of about 2500 cords of wood the length of Aziscoos Lake to the dam, a distance of about 12 miles, where the logs are sluiced into the Megalopa River. At Errol [NH] dam the logs are again sluiced, this time into the Androscoggin River, which transports the logs directly to the mills at Berlin."

Nibroc.

Hendel Photograph Collection, Box 1, Folder 33, Maine Maritime Museum.

THE BROWN Bulletin

QUALITY

VOLUME 2
NUMBER 11

Published monthly for employees and friends of Brown Company
Berlin, New Hampshire

JUNE
1954

Nibroc.

Hendel Photograph Collection, File 1, Folder 37, Maine Maritime Museum.

Circa 1954: *Barbie D*, Camden Shipbuilding's 23' harbor tug, was named for Barbara Dyer, long-term office employee of Camden Shipbuilding and Marine Railway Company/Camden Shipbuilding/Wayfarer Marine, and who has become Camden's official town historian.

Barbie D.

Barbara Dyer Collection, Walsh History Center, Camden Public Library.

1966: *Seasmoke* a 38' steel lobster wholesaler's workboat constructed by Thomaston Steel Works, Thomaston, Maine, for the owners of the Post Lobster Pound in Friendship, Maine.

Seasmoke.
Irving Nevells Collection, Penobscot Marine Museum.

Courier-Gazette article by Irving Nevells:

HEFTY STEEL BOAT IS LAUNCHED
LIKE A SKIFF AT THOMASTON.

THOMASTON, ME—A boat launching in this town is not an unusual occurrence, but the launching of the 38' steel workboat *Seasmoke* on April 30 had a few unusual quirks.

It was no small feat to transport her on the lowbed trailer from the shop of The Thomaston Steel Works through the business section of the town to the Public Landing—she weighs about 13½ tons.

The original plan contemplated the typical tide launching, pushing her down to low water mark and leaving it to the incoming tide to float her. But Eugene Benner of the Steel Works, wanting to hurry the action, conceived the idea of pushing her off into the channel which runs close to the low water mark at this point. It worked.

One last nudge and *Seasmoke* slid gracefully into the channel in a typical outboard launching as Mrs. Ann Post, one of the owners, broke the bottle over her bow.

This sturdy little boat which resembles a small tug will be around the Post Lobster Pound in Friendship. One of her primary functions will be icebreaking, and Merritt Post is sure he will be able to keep the pound free of ice by breaking it up before it is too thick.

Seasmoke is 38' x 12' with 4' draft. The hull is constructed with ¾" carbon steel plates to provide the strength for her icebreaking job. There are four watertight bulkheads of ¾" steel plate. The deck is 3/16" steel plate.

The boat has tugboat power with a 360-hp Caterpillar D 333 diesel, driving a 32" x 20" wheel through a 2:1 reduction gear. Living accommodations included two berths, a head, and Post will include a stove later.

Both boats [see the 22' aluminum lobster boat on the following page 169] were designed by Geerd Hendel, N.A., of Camden. Col. Alfred T. Greene, USMC (ret.) acted as consultant during the building.

California-born Merritt Post and his wife, Ann, who was born in the Philippine Islands, are among the few out-of-staters who have made a success of the lobster business on the Maine coast.

Post says he has owned an island adjacent to the pound for many years. They came for vacations when they were both working in New York State. When the pound came up for sale, they decided to pool their resources and go into the lobster business.

After six years they are well established as lobster dealers.

Seasmoke's (later *Matinicus* and *Tug Alley Too*) story continues:

Hartley Marine Services in Boothbay Harbor bought the *Seasmoke* at auction in 1997 and renamed it *Matinicus*. It was later leased to a construction company and used to tow a 120' by 40' barge with an operating crane. The barge was used along the southern Maine coast and around the Boston area for the same purpose. It also participated in rescue operations at BIW.

Bob and Natalie Hassold purchased *Matinicus* in November of 2001, renamed it *Tug Alley Too*, and secured licensing approval for the tug to provide Portsmouth, New Hampshire, harbor cruises for six passengers.

When the Hassolds' closed their business in 2013, *Tug Alley Too* was sold and is now in private ownership in New Hampshire.

Tug Alley Too *underway in Portsmouth Harbor.*
Robert Hassold Collection.

1966: A Hendel-designed 22' aluminum lobster boat was constructed at Thomaston Steel Works, also for the Post Lobster Pound in Friendship.
Irving Nevells Collection, Penobscot Marine Museum.

MAINE PIONEERS NEW LOBSTER WORKBOAT

What may turn out to be a prototype for inshore lobster workboats was launched by the Thomaston Steel Works in mid-May.

Built to the order of Merritt Post, proprietor of Post's Lobster Pound, Friendship who said he wanted an economical lobster boat with plenty of speed, light in draft and as maintenance free and operationally economical as possible, the boat on her trial run indicated that she would fulfill these requirements.

The 22x7-foot noncorrosive aluminum craft, powered by a 60 horsepower diesel, made in England, with a MerCruiser outboard drive, is believed to be capable of a speed of 30 miles per hour. The inboard-outboard feature permits operation in shallow water and provides more usable room aboard, because lobster traps can be piled on the engine box.

Geerd H. Hendel, Camden marine architect who has designed many larger aluminum workboats, as well as Post's craft, said he believes the little lobsterboat will prove efficient for in-shore summer fishing.

Post, the owner, expressed the hope that many Maine lobstermen would adopt the model. "They'd escape the enormous chores of painting and caulking, and could fish the ledges which boats with fixed propellers must avoid."

Post recently had another unusual craft built by the Thomaston Steel Works. She's the 38-foot steel tugboat, "Seasmoke," designed especially for breaking ice in Lobster Gut, where his pound is located, in Friendship Harbor. The vessel also can break the ice in the pound and force it out through a gate by the propeller's thrust.

Eugene Benner, Thomaston Steel Works manager, declared that the Post boats are the first to be built by his firm, which specializes in all kinds of metal fabrication and construction. "We had such good luck with these boats, we'll be glad to try our hands at building others to special order," he said.

22' aluminum lobster boat with Hendel at the stern.

Hendel Photograph Collection, MS-Box 310, Maine Maritime Museum.

Hendel designed the 1961 steel log-boom tug *William Hilton* and its twin, the 1964 *O.A. Harkness*, for the Great Northern Paper Co. (GNP) in Millinocket, Maine. The *Hilton* and *Harkness* were each 70' long with beams of 20' and were powered by two 250-hp General Motors diesel engines. Two generators powered a 120-volt electrical system and heating system, and the cabins included a pilothouse, dining area, crew's quarters, and head.

The *William Hilton*, named for GNP's Vice President of Engineering and Woodlands Management, was constructed at the Goudy & Stevens yard in East Boothbay in 1961 (hull 192).

SPONSOR AND DAD—Mrs. Judson, Mealy of Sandusky, Ohio, right, greets her father, William Hilton, during formal christening ceremonies which placed the new towboat of the Great Northern Paper Company in service at Chesuncook Lake Saturday. Hilton, for whom the steel vessel is named, served Great Northern in different capacities for nearly 50 years. (NEWS photo by Washburn)

The William Hilton*'s first engineer was Richard "Dick" Thomb, who became her captain one year later.*
Hendel Photograph Collection, Box MS-310, Maine Maritime Museum.

William Hilton *hull and superstructure being transported to Chesuncook Lake.*
GNP Co. photograph from the website of Chesuncook Lake Association.

Columnist John Gould wrote in the *Christian Science Monitor* in 1993 that the *Hilton* was "mostly a power plant. 'Twas said that she was capable of jerking the Empire State Building right over into New Jersey."[107]

The *William Hilton* was based on Chesuncook Lake on the West Branch of the Penobscot River as a towboat for log booms. A crew of six operated the boat, originally on a 24-hour-a-day basis, navigating by compass, searchlight, and skyline landmarks, but towing was later limited to daylight hours. To tow the pulpwood, a double loop or bag" of boom logs each about 26 feet in length and tied together with heavy chains, was created. Towing across the open lakes was relatively simple, but the tug operated at only one mile per hour because a higher speed would have caused the logs to jump over the back edge of the boom. Maneuvering booms through narrow channels could be tricky and limited the number of logs that could be towed at one time. Windstorms also could cause an entire tow to be abandoned until the weather moderated.

The 22-mile long Chesuncook is the third largest lake in Maine, and Gould also reported that "Flint Johnson, who cooked on the *Hilton*, once told me that they towed 50,000 cords of boomed pulpwood into the lake's wind for 17 hours and ended up five miles behind where they started." A testament to the impact of winds on large lakes.

Three years later, a nearly identical steel tug, the *O.A. Harkness*, was also built at the Goudy & Stevens yard (hull 199). It was named for Orris Albert "O.A." Harkness, a Camden/Rockport native who became the GNP Company's Superintendent of Motor Equipment and the "admiral" of its inland fleet.

The *O.A. Harkness*'s home port was at North Twin Lake near Millinocket, and she was used to tow log booms along the West Branch of the Penobscot River across the "Lower Lakes" (Ambejejus Lake, Pemadumcook Lake, North and South Twin Lakes, and Elbow Lake). This was a distance of about

GNP Co. log-boom tug William Hilton.
Special Collections, Raymond H. Fogler Library, Great Northern Paper Company Records, 1889–1992, University of Maine.

Orris Albert "O.A." Harkness.
V.O. Harkness Jr. Collection.

14 miles, and between 2500 and 3000 cords were brought to the Great Northern mill in each tow.[108]

GNP estimated that the *O.A. Harkness* handled about 120,000 cords of pulpwood annually. Figuring 60 to 65 logs per cord, during its working years the *Harkness* had towed enough pulpwood, if placed end to end, to reach about 46,000 miles, equal to a chain of logs circling the earth twice.

The two vessels became available for purchase when GNP phased out its river log drives in 1971. The federal Clean Water Act of 1972, sponsored by Maine's U.S. Senator Edmund Muskie, was ushered into law by the U.S. Congress. Muskie, who was from the paper mill town of Rumford, knew what the logging and paper mills added to the Maine economy, but he deplored what they did to the rivers. It was partly the environmental impact of the industrial process wastewater discharges from the paper mills along the river that brought the Clean Water Act into effect. The log drives contributed to environmental impacts as bark from the logs settled on the river bottoms causing pollution and removing habitat for fish spawning. This drastically reduced populations of both fresh- and saltwater species. Logs in the rivers disrupted boaters, fisher-

GNP Co. log-boom tug O.A. Harkness.
Special Collections, Great Northern Paper Company Records, 1889–1992, Raymond H. Fogler Library, University of Maine.

men, and canoeists. Paper manufacturers made the adjustment to transporting logs by truck, building hundreds of miles of logging roads and constructing some new mills.[109]

The *William Hilton* was bought by Captain Eliot Winslow of Southport Island, Maine, around 1972. The vessel was trucked down to Bath Iron Works, sandblasted, and put back together. Hendel was given the design work for updating the vessel. In an early 2017 conversation with the author, David Winslow of Winslow Marine in Falmouth confirmed the *Hilton*'s purchase by his father. He recalled that the Murphy diesel engines soon failed, and they were replaced with Detroit Diesel 12-71's. Winslow Marine renamed her *Argonaut* and secured Coast Guard approval to carry 69 passengers for fishing and sightseeing. Winslow said she was not really effective for fishing parties and wasn't well-designed for use as a harbor tug, although it was used as one. Winslow Marine sold the boat to Gibson Construction of Portland around 1985, and it was owned by that firm for at least eight years. John Gould reported that as of 1993, she was a tug in Portland Harbor, but the actual trail of subsequent ownership and location is uncertain at this point.

The *O.A. Harkness* was purchased by the Maine Department of Sea and Shore Fisheries in 1972 and was judged to be well-suited to the needs of the department. The vessel was transported by truck to Bangor, where it was launched into the Penobscot River to make its trip to Boothbay Harbor, where it was based.

It was later sold by the State to Harold MacQuinn, Inc., a construction firm based in Bar Harbor, and used to tow construction company barges along the Maine coast. It was engaged in such a tow when it began taking on water on January 16, 1992, just two miles south of Matinicus Island. The leak seemed to be coming from a snapped rudder post.

The *O.A. Harkness* sank after dark in seas with eight-foot swells and a wind-chill air temperature of –54 degrees. When the three men wearing immersion survival suits abandoned the *Harkness,* one of them had grabbed a flashlight.

Lobstermen were listening to the marine radio on Matinicus Island, and Vance Bunker and two friends set out in Vance's lobster boat to attempt a rescue. A Coast Guard boat was also dispatched from the Rockland Coast Guard Station. The three crew members were in the water together clinging to a bit of wreckage, and the flashlight was turned on. The flashlight beam was quickly seen through the sea smoke by those on the lobster boat, and the three *Harkness* crew, Captain Rudi Musetti, Mate Arthur Stevens, and crewman Duane Cleaves, were rescued and taken to Matinicus Island.[110]

Hendel's friend and Maine author John Gould wrote, in an article entitled "Recipes for Towboats and Sweet-and-Sour" in the *Christian Science Monitor* edition of May 7, 1993, that he thought the *William Hilton* to be Hendel's finest design effort. In it, he humorously noted, however, that "There has got to be something bizarre in fetching the talent of the Hanseatic shipbuilders to America to make a towboat on Maine's Chesuncook Lake. But here was Geerd in Maine, and the order was a challenge. As you can suppose, certain technical adjustments were necessary to accommodate deep-water conditions to fresh water, and Geerd spent long hours at his drafting board." The *Hilton* and *Harkness* both served their functions admirably.

Tug William Hilton *towing a log boom on Chesuncook Lake.*
Special Collections, Great Northern Paper Company Records, 1889–1992, Raymond H. Fogler Library, University of Maine.

1989: 26' aluminum *Welcome* was designed for the Town of Camden, Maine, for use as its harbor tug operated by the harbormaster. Built by John Luke at the Paul E. Luke boatyard in East Boothbay, its dimensions are 25'6" LOA, 9'10" beam, and 17" draft. The 130-hp diesel provides a top speed of 12 knots. Among its unique features are dive stairs, an amber strobe light, and a water cannon hookup for firefighting.

1975: *Alton A. II,* a 44' LOA steel tug hull built by Gerry Libby of Damariscotta Boats, Inc., in Rockland, Maine, for the marine construction and dredging firm Prock Marine, Inc. Libby, who had owned and operated the Melvin Heights Boat Shop in Camden during the 1960s, owned Damariscotta Boats in the 1970s. The yard specialized in constructing aluminum and steel commercial vessels. Records show that Libby built the *Alton A. II*'s hull, and Prock Marine installed the tug's engines, generators, and other equipment in their Rockland yard. The tug was launched in 1982.

Welcome.
Town of Camden, Maine.

Prock Marine tug Alton A. II.
Prock Marine, Inc., Collection.

CHAPTER TWELVE

Paper Company "Boom Jumpers"

O.A. Harkness, on the suggestion of GNP's Vice President Fred A. Gilbert, developed the smaller "boom jumper" boats to fill the need for a boat that could operate when filled with four-foot-long pulpwood held together by floating log booms that needed to be crossed. The boom jumpers were used "primarily to do errands and light work around booming-out places and in bringing up and placing the slack boom. When it is necessary to cross a boom and get inside, the boom jumper is headed directly upon it. The engineer opens the throttle and the bow hits the boom with a splash; there is a moment's hesitation, then the boat rides smoothly on the forefoot and slides easily over."[111]

To describe how boom jumpers crossed log booms, columnist Lew Dietz, in the May 1967 *National Fisherman* wrote: "It took a special boat and a special man for the tricky job of jumping booms. Harkness designed the boats for the job, and in due time there were men with the cowboy knack of handling them. When it was necessary to get inside a boom, the boat was headed directly at the barrier, throttle wide open. The trick was to hit the boom with the bow up so that the boat would rise smoothly on her forefoot and slide over the obstacle in one splashing leap."

Prior to 1950, GNP boom jumpers were designed in-house by O.A. Harkness, GNP Co.'s Superintendent of Motor Equipment and the "admiral" of GNP's inland fleet. Upon Harkness's retirement in 1950, however, the 1951 *GNP Co. No. 45* and later "log-boom jumper" tugs were designed by Hendel and fabricated of steel. These tugs were typically 35' LOA, with a beam of 9', a draft of 4', and were powered by 100-hp Lathrop diesels. The first series of GNP steel boom jumpers constructed after 1951 (*GNP Co. No. 45*, *No. 46*, and *No. 49*) were constructed at the General Foods Corp. shipyard in Rockland, Maine, and transported by tractor-trailer 200 miles north to work at Chesuncook Lake on the North Branch of the Penobscot River.

1952 steel "log-boom jumper" tug *GNP Co. No. 46* was launched at the General Foods Corp. shipyard in Rockland on April 18, 1952, to work at Chesuncook Lake.

GNP Co. No. 49, a 35' steel log-boom jumper/tug, was launched at the General Foods Corp. shipyard in Rockland on April 30, 1954, to work at Chesuncook Lake on the West Branch of the Penobscot River. She was christened by Mrs. John Maines.

GNP Co. No. 45, *built in 1951, steel hull, in Greenville.*
Chesuncook Village Association.

GNP Co. No. 46 *readied for launching.*
Special Collections, Great Northern Paper Company Records, 1889–1992, Raymond H. Fogler Library, University of Maine.

GNP Co. No. 49 *and* Whitecap *(see text below), probably in Rockland Harbor.*
Geerd Hendel Collection, courtesy of his stepdaughter, Brigitte M. Gefrich.

Hendel's design for the 1954 35' *Whitecap* was very similar to that of *GNP Co. No. 49*, but she had a slightly deeper draft. Power was a 170-hp GM diesel engine. She was built for the Hollingsworth and Whitney Paper Company, which was based in Waterville and Winslow, Maine. *Whitecap* was the last vessel built by the General Foods Corp. at their Rockland yard.

The descriptive information for vessels *GNP Co. No. 45, No. 46*, and *No. 49*, and *Whitecap* is from Bertram Snow's book *The Main Beam*, published by the Rockland, Maine, Historical Society.[112]

An April 4, 1957, article in the *Boothbay Register* reported that GNP Co. had ordered three steel boom jumpers to be fabricated by the Hodgdon Bros./Goudy & Stevens partnership in East Boothbay. The 35' steel boom jumper shown on page 180 (hull 132) was built in 1957, and the two other boom jumpers of similar design are thought by the author to be the 35' steel vessel *GNP Co. No. 57* (hull 143) built in 1956–57 and a 28' steel vessel (hull 146) built in 1957.

35' steel boom jumper built in 1957 by the Hodgdon Bros./Goudy & Stevens partnership.
Boothbay Region Historical Society.

A steel boom jumper at Goudy & Stevens, East Boothbay, circa 1957, being loaded for transport by a GNP Co. truck and lowbed trailer. The porthole style windows seem to be a characteristic feature of the Goudy & Stevens boom jumpers.
Boothbay Region Historical Society.

The newspaper article reported the following:

> Hodgdon Bros, Goudy and Stevens, East Boothbay Shipyard, recently completed work and trial runs on one of three 35-foot steel tugs for Great Northern Paper Company. The boat has a beam of 10'2"; draft of 5'2"; and is powered by a GM 6-71 Detroit Diesel, 4:5:1 reduction. The five-blade propeller measures 48", and the boat has a capacity of 5000-pound towing on bitt. There is a hydraulic throttle and clutch and Crowell hydraulic steering controls.
>
> Hydraulic reverse and steering make for easy handling and maneuverability. In fact, an important feature of this boat is the Crowell Steerer, as it enables the wheel to be left on course, unattended, while some other operation on the boat is carried out. This factor eliminates the need for an extra hand.
>
> These boats have proved to be remarkably successful for the work they were designed to do, as the repeated orders for more have amply demonstrated.
>
> Similar models have been built at the yard previous to this group of three, all of them for the Great Northern. These tugs are designed for rough service in handling logs. They are all welded steel construction and reinforced to withstand boom log jumping and acres of pulpwood. They can handle up to 5 acres of floating logs.

This boom jumper, *GNP Co. No. 57* (hull 143) of Goudy & Stevens construction, appears to be set up for a demonstration or boat parade with the KEEP MAINE GREEN & CLEAN sign on the cabin and three loggers on the bow (two holding a bucksaw and one holding a chain saw) staging a cutting pose over a demonstration log.

In 1963, a 30' steel workboat was built by Goudy & Stevens for GNP as a "boom tender." The photograph below was included in the June 1963 edition of the *Maine Coast Fisherman* in an article titled "Boothbay Area Famed for Boatbuilding Yards." The photo's caption notes that the hull was designed with "tunnel drive," which would protect the propeller from its work environment of floating four-foot pulpwood logs.

GNP Co. No. 57.
Moosehead Historical Society Collection.

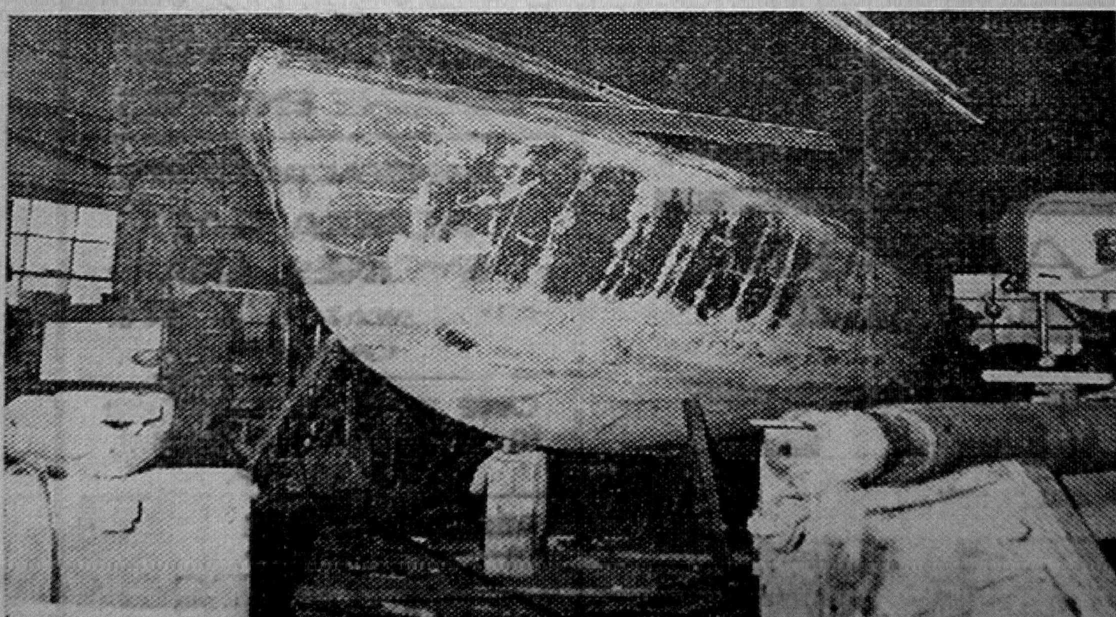

STEEL WORKBOAT, a 30-footer with tunnel stern to be used on big wilderness lakes as a boom tender by the Great Northern Paper Co., comes to life in Goudy & Stevens shop. Power for boat will be a GM 4-71 diesel.

June, 1963 NATIONAL FISHERMAN

CHAPTER THIRTEEN

Collaboration with Naval Architects, Designers, and Boatbuilders

Hendel occasionally collaborated with fellow naval architects. To clearly separate that work from work that was directly his own, this chapter names these fellow professionals, generally following the progression of his career, and describes those interactions around specific watercraft.

NAVAL ARCHITECTS

W. Starling Burgess

In his design and drafting work for naval architect W. Starling Burgess in the last half of the 1930s, there was ample opportunity for him to incorporate his own design refinements to those of Burgess. Burgess's creative, energetic, and daring spirit, combined with incredible mathematical acumen, led to the development of revolutionary hull design concepts. Hendel was always careful to credit Burgess with establishing the design details from which he drafted plans. As noted in Chapter 5, during Hendel's oral 1988 interview archived at the Mystic Museum Research Library, his interviewers focused on the Yankee One Design, seeking to learn from Hendel whether or not he had taken a lead role in the racing sloop's innovative hull design. Hendel demurred, however, saying that the YOD design ideas came from Burgess, and his role had only been to translate them into design drawings.

Albert E. Condon

Naval architect Albert Condon designed two draggers, the *Pelican* and the *St. George,* for the Snow Shipyards, Inc., in Rockland. The lines for the *St. George* were completed on January 2, 1939. In the spring of 1939, Condon submitted his resignation as the Snow yard's naval architect and yard superintendent effective in early October 1939, and Hendel was then hired in the spring of 1939 to succeed him as the yard's naval architect. Ernest Gamage succeeded Condon as yard superintendent.

1939 dragger St. George, *designed by naval architect Albert Condon.*
Rockland Historical Society, Rockland, Maine.

The wood-hulled dragger *Pelican* (later *Porpoise*) was 79' LOA, and was built for Carl Beckman of New Bedford, Massachusetts, which was also to be her home port. She was launched in September 1939.

Construction of the 110' wood-hulled *St. George* began in October 1939 and was completed in February 1940.[113]

In his 1988 interview, Hendel noted that he had known Albert Condon on a professional basis. Condon designed 153 vessels, the plans of which are archived in the Daniel S. Gregory Ships Plans Library at Mystic Seaport. Although Condon did design some commercial cargo carriers and a few yachts, between 1930 and the early 1960s he principally designed eastern-rig fishing draggers and scallopers of 60' to 110' LOA. Many were built at the Newbert & Wallace and Gray Boatyards in Thomaston, Maine. Condon considered Thomaston his home and retired there in 1950.

Hendel's temporary assignment with the Snow Shipyard ended in 1940, when, to support the U.S. Navy's World War II efforts, he took a job at the William Robinson Shipyard in Ipswich, Massachusetts.

E. Farnham Butler

The Maine Coast "36" yawl (ca. 1945) was an E. Farnham Butler (1909–2006) design with Hendel serving as consulting naval architect. In 1943, Butler established the Mount Desert Boatworks in Northeast

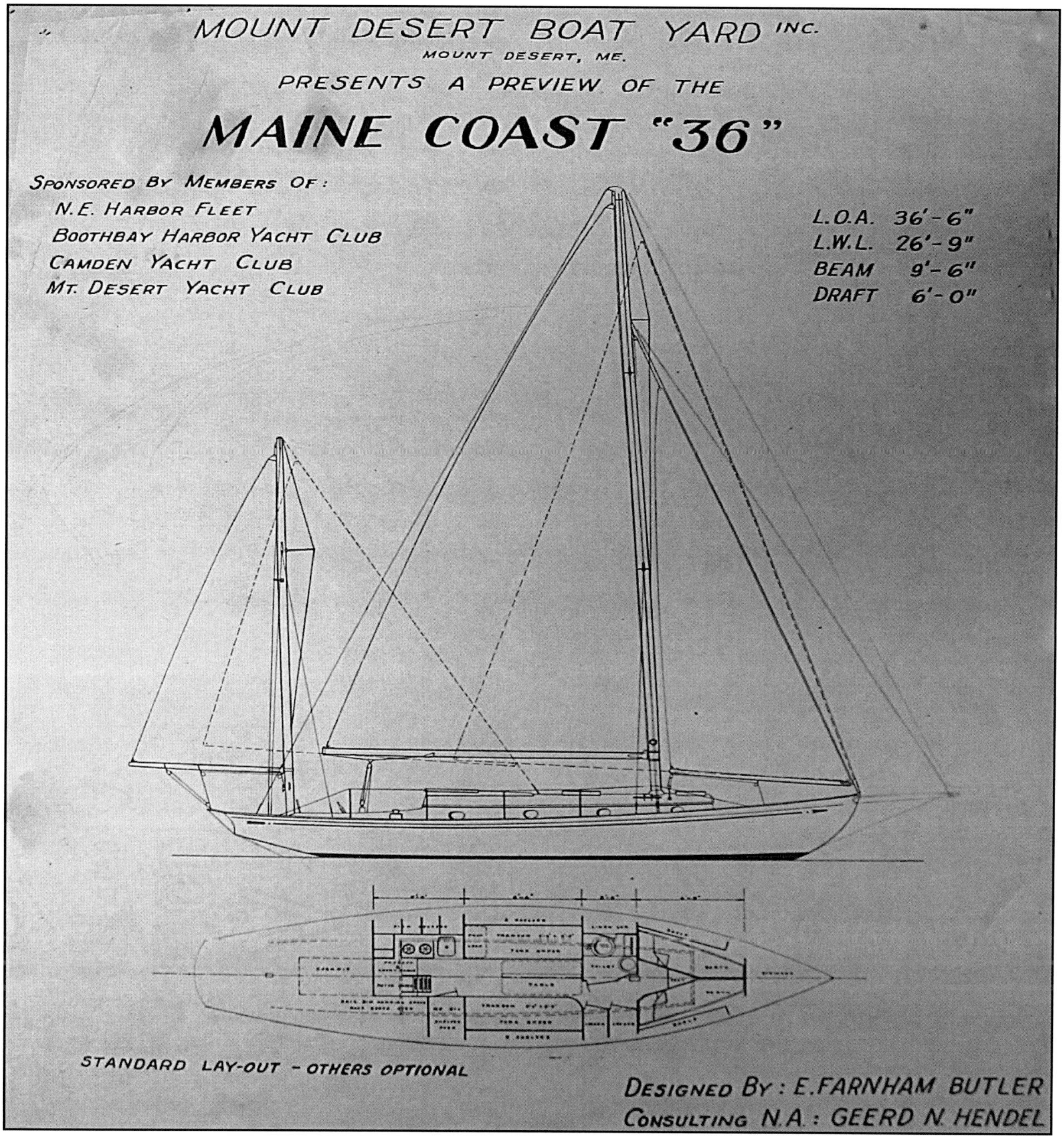

Hendel is credited as "Consulting N.A." on this informational flyer.
Hendel Photograph Collection, Box MS-310, Maine Maritime Museum.

Harbor on Maine's Mt. Desert Island. At the close of WWII, the yard prepared for the anticipated post-war boom, which was thought to favor a stock, series-built production boat.

The handsome, traditionally designed, Maine Coast "36" was Butler's first attempt to prepare for that market. Maynard Bray recalled that three of these yawls were built: 1945 *Snowflake* (for Mr.

Guthrie), 1947 *Torna* (for Charles Cook), and *Suliko*, which was acquired in 1972 by Charles and Carol McLane, who sailed from Brooklin, Maine. Charles is the author of several books on Maine islands, including *Islands of the Mid-Maine Coast* (The Kennebec River Press, Woolwich, Maine, 1982). These boats were also known as "Snowflake Yawls," honoring the first of the series.

Butler then moved on to design and build, during the 1950s and 1960s, the light displacement 26', 27.5', 31', 36' and 37' Controversy line of small sailing yachts. The first Controversy (28.1') was built in 1951 and displaced a very light 6900 lbs. The Controversy 36 was advertised as costing half as much as a conventional boat of the time. During the decade of the 1950s, Butler's 24' Amphibi-ette and 25.4' Amphibicon were ahead of their time with their reverse sheer and glued strip-plank construction, light displacement, and trailerability.[114]

Howard I. Chapelle

1963 *Old Baldy* was a 25' Friendship sloop built by Jim Rockefeller in his Bald Mountain Boat Works. She was a replica of the 1914 Friendship sloop *Pemaquid*. Launched at Rockport Marine, she was 25' LOA with a beam of 8'7", carried a sail area of 432 square feet, and was powered by a 25-hp Universal Atomic 4.

1965 Sazerac, *the 25' sister vessel to* Old Baldy, *was built for Georges McKinnon at the Bald Mountain Boat Works.* James Rockefeller Jr. Collection.

Master boatbuilder Malcolm Brewer with a model of Pemaquid.
James Rockefeller Jr. Collection.

Detailed plans for *Pemaquid* were taken from the 1935 lines drawn by Howard I. Chapelle and redrawn by Jay Hanna of Rockport, Maine, with information developed for Bald Mountain Boat Works by Geerd Hendel, Malcolm Brewer, master boatbuilder, and Henry Bohndell, sailmaker.[11]

Hendel prepared the construction drawings, keel detail, and sail plan for *Old Baldy*, per his invoice to Bald Mountain Boat Works.

Jarvis Newman, at his Southwest Harbor, Maine, yard created the mold for the fiberglass versions of the *Pemaquid* in 1965 from *Old Baldy*, giving Jim Rockefeller free storage at his shop in Southwest Harbor in return for letting him take a mold off the boat.

Murray Peterson

MIT-trained naval architect Murray Peterson (1908–1974) designed traditional Maine sloops and schooner yachts reminiscent of the historic past to serve as family cruisers. In 1929, Peterson went to work in

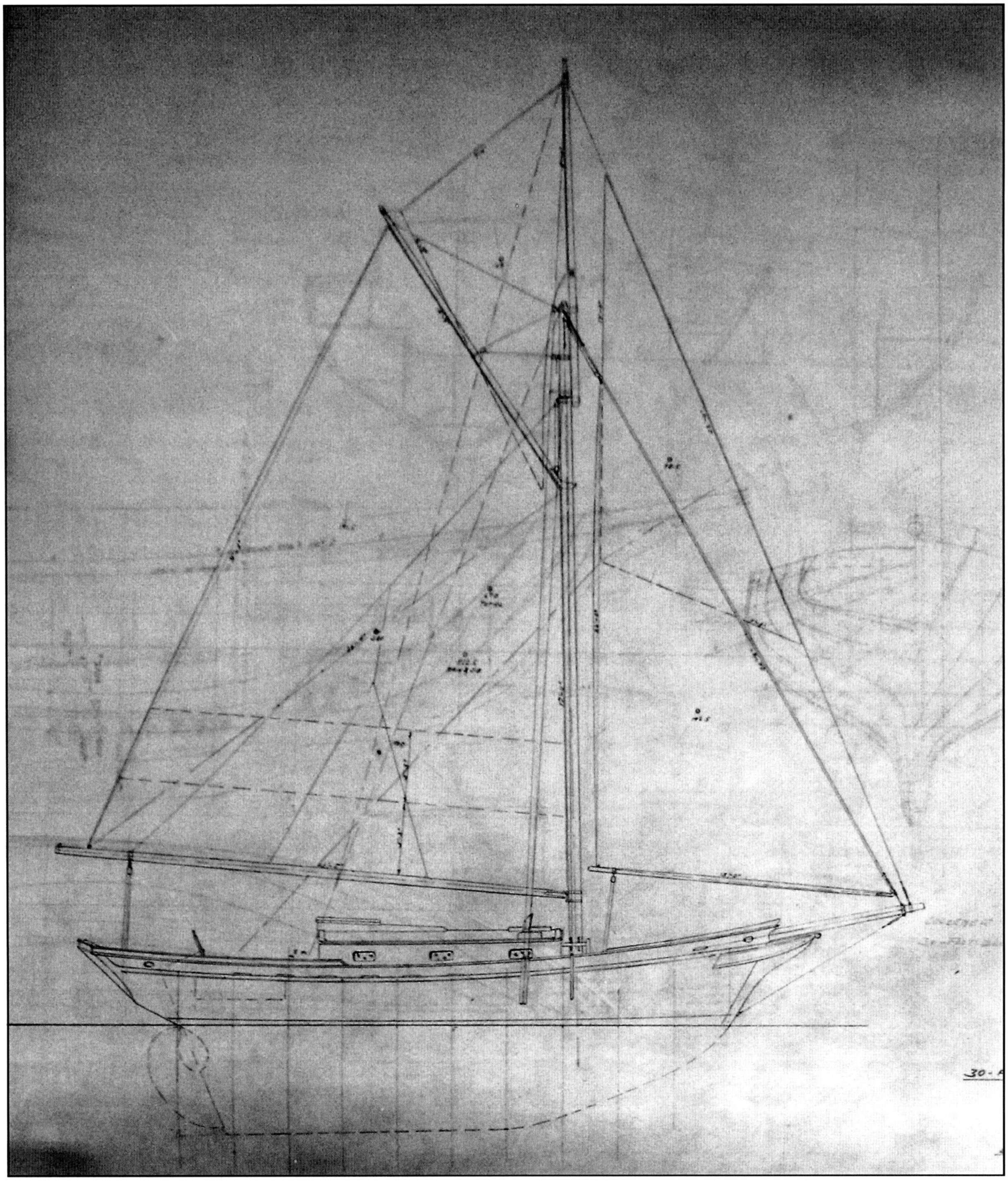

Heritage *sail plan.*

Hendel Photograph Collection, Box 1 List, PL 19 98.106.4, Maine Maritime Museum.

John Alden's design office and established his own design office in Maine's midcoast region in 1933. Notable schooners of Peterson's design later in his career include the 36' *North Star*, the 42' *Silver Heels*, and the 42' *Serenity*. All were schooners built by Malcolm Brewer and the crew at Camden Shipbuilding in 1962–63.

The 30' auxiliary sloop *Heritage* was designed by Peterson and built by master boatbuilder Elmer Collemer in 1963. She appears to be the only Peterson-designed boat in which Hendel had a design or drafting role, probably assisting Collemer directly as he built the vessel. Detailed plans for *Heritage* are in Hendel's archival collection at the Maine Maritime Museum. The drawing shows the sail plan for the vessel, but the Museum file also contains a complete set of the vessel's construction plans with no attribution as to who drew them.

Heritage.

"Historic Camden-Built Sloop Returns to Midcoast," The Free Press, *Saturday, July 5, 2014.*

DESIGNERS AND MASTER BUILDERS

One great uncertainty of Hendel's career is the extent to which he provided supportive design and drafting assistance to master boatbuilders and designers, especially in the Camden area. His deep respect for their skills and craft, combined with his innate modesty, led him to assist in providing drafting and perhaps design assistance to support their work, but he sought no credit for it.

Following are the designers and builders for whom he either assisted with designs or provided drafting services.

Jim Stevens: designer, master loftman, and boatbuilder

At the age of 25, Jim Stevens was hired from the Goudy & Stevens Yard in East Boothbay to be manager of the lofting department by the new Camden Shipbuilding Company in 1941.

While waiting for the U.S Navy's shipbuilding work to restart after the launching of minesweepers AMc82 and AMc83 in early 1942, Stevens lofted a 40' wooden fishing dragger to keep the yard crew busy. The *C.W.W.* was launched in March of 1942. In an interview with Boothbay Region Historical Society volunteer Barbara Rumsey on May 2 and May 16, 2002, Stevens said, "I designed her on the mold loft floor. The initials were for the three fellows who had her built: Carver, Whitney, and Wallace. Only when they came to see her, I'd built her so heavy that they called her 'Carver's Wooden World.' We just got her done and delivered to them and the Navy drove [*sic*] Camden—Hendel, Peterson, etc." [that is, to restart their Navy work]. *C.W.W.* was later acquired by the U.S. Navy. Hendel's plan files at the Maine Maritime Museum contain some detailed drawings for the vessel.

Malcolm Brewer: designer and master builder

Hendel's drafting work at Camden Shipbuilding was supervised in part by its master builder, Malcolm Brewer, who was also a boat designer in his own right. Hendel first met Brewer at Hodgdon Bros. in East Boothbay in 1932, when Hendel was hired by the nearby Rice Brothers Corporation to work on a fireboat being built for the city of Portland. Hendel observed at the time that "Malcolm's strong point was the quality of his work. He could explain to the men, took his time, was patient with them."[116]

In *Grog Ho*, Barbara Dyer writes that Malcolm Brewer was born in 1899 in East Boothbay, Maine, and became a joiner in the Hodgdon Bros. yard. About 1939, he left Hodgdon and established his own boatbuilding shop in East Boothbay, but when WWII began and materials became scarce, he then worked for BIW. He came to Camden Shipbuilding in about 1941 to lead the construction process for twelve salvage rescue tugs (ATR's) for the U.S. Navy.[117]

> His [Brewer's] workmen respected him, appreciated his great abilities, and knew that he was "boss" in his quiet and moderate way. His friends and workmen can still picture him in his usual attire—felt hat, khaki shirt and pants, and elastic armlets. In cold weather, it was topped off by a Navy pea coat. He could follow any Naval Architect's design or owner changes, making suggestions for improvement which were readily acceptable. Geerd Hendel and Murray Peterson were the two men he worked with most often, and their respect of skills [was] mutual. His workmen were ever learning, and some became very highly skilled craftsmen under his guidance. He listened to many of their suggestions, and he also learned. Malcolm was from the "old school," where workmen were expected to be punctual, meticulous, and had to do the work with absolute precision.[118]

After the war, Camden Shipbuilding turned to constructing custom pleasure yachts built under Brewer's supervision.

In 1945, Camden Shipbuilding manager William E. "Pete" Peterson recalled the discussion over the nature of the first yacht the yard would build as it transitioned from building wartime vessels to custom pleasure boats: "Malcolm produced a half model of a 40' sport fisherman he had 'dreamed up' and suggested that such a craft would provide a good training ground for the workmen who has been 'spoiled' by the war program, which called for speed and strong construction with little or no requirement for fine finish or elegant detail."[119]

It was agreed to start construction on two of the 40' sport fishermen, and Brewer lofted the *Andrea B* and *Marbaru* from his own half models.

1946: *Andrea B*, a 40' sport fisherman built for Earl Boardman by Camden Shipbuilding (hull Y-1), was based on a design from Camden Shipbuilding's master boatbuilder Malcolm Brewer.

Andrea B.

Barbara Dyer Collection, Walsh History Center, Camden Public Library.

Hendel's role in the design of these two boats isn't entirely certain, but probably included the cabin arrangement. The complete design plans for both boats are included in his designs archived at the Maine Maritime Museum. Given Brewer's supervisory role in Camden Shipbuilding's management structure and his positive working relationship with Hendel, it's logical that he would assign drafting and other design work to Hendel, who was Camden Shipbuilding's in-house naval architect from 1941–45.

1946: Marbaru *(later* Rag Doll*), a 40' sport fisherman built for Clinton Lunt by Camden Shipbuilding (hull Y-2). She is a sister vessel to the* Andrea B.

Bayport Yachts, Chester, Maryland.

In 1945, Hendel established his own independent naval architecture practice in Camden but continued to design custom boats for Camden Shipbuilding, in addition to those for other clients and yards.

Brewer himself designed many boats, often from half-hull models he created.

1949: Snapper, *a 35' cruiser built for Leverett Saltonstall by Camden Shipbuilding, was powered by a 120-hp Mercruiser engine and outdrive. She was designed by Malcolm Brewer, with Hendel providing drafting services. This image is from Hendel's personal photo album.*

Geerd Hendel Collection, courtesy of his stepdaughter, Brigitte Gefrich.

1968: Felicity *is a 33' (sparred length of 46') clipper-bowed schooner designed by Hendel and Elmer Collemer for Pierre Manigault. Elmer Collemer built her at his one-man Camden shop, but died before completing the vessel. She was finished by Southwest Boat in Southwest Harbor, Maine.*

Registry of Wooden Boats, *WoodenBoat* magazine.

Elmer Collemer: master boatbuilder

The 30' auxiliary sloop *Heritage,* noted earlier in this chapter, was designed by Murray Peterson and built by master boatbuilder Elmer Collemer in 1963 at his shop on Cross Street in Camden. Peterson's design reflects that of the Friendship sloop, a traditional Maine working vessel. Detailed plans for *Heritage* are contained in Hendel's archival collection at the Maine Maritime Museum and contain drafting attributions to him.

Heritage is now owned by Captain Jeff Beck, who operates her out of Camden harbor for day charters for up to six passengers.

Author and boatyard owner Dan MacNaughton noted to the author his impression that Hendel may have designed *Felicity,* the last boat built by Elmer Collemer, which he's certain was built for Oaksman Smiley of Bangor, Maine. He commented that *Felicity*'s design is usually credited to Murray Peterson, but it could have been Hendel's subtle aesthetic reworking of a Peterson design; he considered the design "exquisite."

1951: Andiamo *("Let's Go") is a 38' auxiliary sloop built by Camden-based boatbuilder Elmer Collemer for William K. Wipple of Pulpit Harbor, North Haven Island, Maine.* Andiamo *is the largest boat built by Collemer. This image is from Hendel's personal photo album, so it's likely that he assisted Collemer with the design drawings for the sloop.*

Geerd Hendel Collection, courtesy of his stepdaughter Brigitte M. Gefrich.

Epilogue

Whether one attributes their lifetime choices to fate, destiny, or the Almighty, for Hendel his many years as a naval architect demonstrate the fortunate circumstance of aligning art and design, personality, friends, and geographic location to optimize his chosen profession and lifestyle. He concludes his 1989 résumé with this: "I found my work rewarding, not a burden, met interesting people and also made friends doing it." Really, what more could one ask from their chosen career?

Geerd Hendel designed an amazing array of vessels over his career as a naval architect, a testament to his design skills and love of his profession. As a practical matter of geography, the range of vessels he designed responded to the variety of commissions that were available to him in Maine and from owners along the northeast coast of the United States.

W. Starling Burgess

Hendel's working relationship with Burgess began in 1934 with the aluminum *Tinavette* project and continued through 1937 with the design development for *Ranger*. Burgess then developed design concepts for an aluminum destroyer for the U.S. Navy in 1938 and early 1939, which Hendel translated into design plans.

Hendel's personal interrelationships with Starling Burgess deserve mention. Although details of their interactions are few, the very fact of his extended personal and professional relationship with Burgess over the decade of the 1930s supports the fact of their design process compatibility. Burgess's creative intellect and massive ego would have driven most marine draftsmen mad, but Hendel's own intellect and his steady nature, from all references in the historical record, created a design interrelationship that flourished. In a 1994 letter to a student interested in Hendel's career as a naval architect, Hendel wrote, "I was very fortunate to have had the opportunity to work with Mr. Burgess."[120]

In Hendel's April 20, 1939, letter to his mother, however, he expressed "issues" with Burgess. (The letter is written in German, and Regine Harding of Rockland, Maine, provided a general translation of page 2 of the letter on October 21, 2018):

> Hendel has some serious issues with the relationship between Burgess and himself, and that's so disappointing. Hendel feels he should have stopped working for Burgess. Burgess has a very advanced knowledge of mathematics, but there is very little in him to be admired otherwise. His character is of the lowest sort. Hendel hasn't worked for him in several months, and Burgess won't have work for him for several more months. Burgess has asked Hendel to wait for the arrival of more work. Burgess owes Hendel $300, which he fears he might not receive if he leaves Burgess's employment.

Hendel's reference to Burgess's character as being of the "lowest sort" probably stems from the fact that Burgess was married five times. He wed his fourth wife, Nanni Dale Biddle, in 1933. She was the principal financial backer of the futuristic Dymaxion Car project, on which Burgess worked with Buckminster Fuller. In 1937, Burgess had hired Marjorie Young (1913–2005), a talented designer 25 years his junior, as secretary and designer and began an extramarital affair with her while still married to Nanni, whom Hendel must have known well.

Hendel continues the letter with his musings about what the future holds for him at that point in time:

> Hendel wants to start his own boatyard specializing in small boat production and aluminum hulls. ALCOA won't support that effort financially but would support him with technical knowledge.
>
> The United States is now in a financial depression with the prospect of war on the horizon. Hendel has a contract for a 20' aluminum boat if there is no war. A shipyard in New York wants to hire him to design large ships, and it would be easy for him to find work there. Hendel doesn't want to work for Burgess, who is moving to New York. He feels at home in Maine and has many friends here. But if staying in Maine is not meant to be, he may have to move to take the New York shipyard position and work to develop his own independent career path in the future. It's more difficult to leave Maine when summer weather is coming, and Hendel doesn't want to move into the city. Any decision about a job change will have to be made soon, and it's a difficult decision for him.

Yacht Design Aesthetic

Hendel's design aesthetic, especially for super-yachts, sailboats, and pleasure boats, has a timeless, classic character that has carried a number of those vessels through five or six decades of continuous use due to the high regard of their owners for their graceful lines and quality appointments. Super-yachts such as *Vedersein*, *Claybeth*, and *Pearl Necklace* are prime examples, but there are many others.

Smaller yachts such as *Sooner/Voyager* have weathered the changing preferences of yacht owners and remain in beloved active use.

None of Hendel's designs are more respected and revered than the Boothbay Harbor One Design, and the Boothbay Harbor One Design Association (BHODA) will ensure that these beautiful and classic racing sloops receive careful preservation and are engaged in active use for the foreseeable future.

In the 1988 oral interview with Maynard Bray, Joel White, and Fred Calabretta, Hendel was asked which of his designs were his favorites. He first responded that he liked to design smaller boats, as opposed to large oceangoing commercial vessels, because he could take control of every aspect of the designing and construction process. Then he named the dragger *Pocahontas,* the sport fisherman *Porpoise,* and the 110' yacht *Vedersein* as three favorites. In response to a question about the design differences between *Porpoise* and *Pearl Necklace,* Hendel noted that the latter has a longer hull and more freeboard.

Joel White's glowing review of *Porpoise*'s design bears repeating:

> *Porpoise*'s hull is nearly perfect; I can think of no change that would make her better. Her superstructure is nearly so. Geerd Hendel, who designed this paragon, managed to blend the two elements, hull and superstructure, into a unified whole that delights the eye, or at least my eye....But it is not her bowsprit or mast that makes her design so unique or distinctive—remove them and she would look better, if anything. The long, continuous flow of the sheerline, starting from a relatively low freeboard at the after end of the superstructure and rising steadily to the flaring clipper bow profile, is one element of the design that gladdens the eye. The low freeboard is made to appear even lower by the continuous guardrail at the deck edge.
>
> The other element that contributes so much to the overall impression of the boat is the low, short deckhouse structure. This superstructure is very much concentrated amidship[s], leaving a good deal of hull showing alone at the bow and the stern. The midships handrail, with a varnished teak rail cap set on bronze stanchions, and a canvas weather cloth laced between the handrail and the cap rail, was designed to give one a feeling of security while walking the side decks. It also does much to lower the apparent height of the deckhouse. I realize that I have spent a lot of words analyzing the appearance of this boat, but beauty is so rare and nowadays so neglected in the design process that it seems worthwhile to consider.

White's comments would certainly apply to many of Hendel's designs.

His design for *Vedersein* has been similarly described as having a relatively small superstructure for her long, lean hull, plus modest streamlining and a timeless quality, all common elements of Geerd Hendel's design work. *Vedersein* has been lauded as one of the most handsome and well-proportioned yachts ever built, and her mahogany hull may make her the largest wood-hulled motor yacht ever constructed until recent years.

Hendel also regarded his designs for *Gone Away* and *Alphard* as especially beautiful. The two sailboats are nearly identical in hull shape and interior layout, but *Alphard* features a raised deck to provide more headroom.

Hendel's treatment of superstructure and cabin windows are a distinctive visual element of his designs; many of his cruisers and some sailboats have carefully graduated window sizes that reflect the vessels' sheers and cabin-top profiles. And on super-yachts, his careful radius on the leading and trailing windows remains a distinctive feature in his designs.

Llewellyn Holman wrote in *No Ordinary Being* (p. 313), in the context of the *Alumette* project, that Hendel was Burgess's "gifted chief draftsman and design associate." He further recognized the significance of Hendel's designs (p. 320), saying, "Like other boats of Hendel's design, *Whistler* was bold in concept, pleasing to the eye, and blessed with a long, successful life."

Hendel commented during his oral interview that his design preference was to always place the hull's point of maximum beam forward of the amidship point, which functionally always gave a boat more stability.

Hendel's stepson-in-law, George Gefrich, recalled that Geerd told him he was able to get excellent speed with his sport fisherman designs, regardless of size, partly by balancing their weight distribution to operate with a fairly level profile. Photographs of Hendel-designed sport fishermen when under way do indeed show a level profile that made the use of trim tabs unnecessary.

Design Materials

Over the length of his career, Hendel's preferred hull material was the proven and high-quality combination of mahogany over oak frames, topped with teak decks, which yielded seaworthy and long-lasting hulls for both motor-powered and sailing craft.

His pioneering use of aluminum in pleasure boats in the late 1930s, however, was remarkable. At a time when great uncertainty surrounded the efficacy of aluminum in saltwater environments, he was steadfast and confident in his knowledge that aluminum alloy and saltwater environments were entirely compatible. When paint was carefully and regularly applied and properly maintained, and electrolysis prevented by the use of buffering materials, aluminum alloy qualified as a superior boatbuilding material.

His work in designing aluminum-alloy boats for saltwater environments, beginning with *Whistler*, also speaks to his adventurous spirit in seeking new and better materials and technologies for boat construction, all to create better-quality boats that competed favorably with the designs of his contemporaries.

Hendel designed the Robinhood 42 (see Chapter 6) to be of fiberglass (called "reinforced plastic" at the time), but that appears to be the only vessel Hendel specified for such construction. And his 1988 interview indicates that none of those vessels were ever built.

Working Boats

Although reviewers naturally focus on the exterior design and interior layout of pleasure craft, we are compelled to marvel that models of two of his fishing trawlers are displayed in nationally prominent museums: *Pocahontas* at the Maine Maritime Museum and *Albatross* in the Smithsonian's marine history collection. These two vessels must surely be classic representatives of their type and function.

Hendel was also completely comfortable with designing steel hulls, as demonstrated by the many steel-hulled towboats, boom jumpers, and other workboats from his design boards.

His history of designing boom jumpers and logging-boom towboats in Maine's forested north country extended from 1951 to about 1972. In the early 1970s, river-based log-boom drives were ended by state and federal legislation for reasons of providing environmental protection to fresh waters. Repeated tug and boom-jumper orders from paper companies and hydroelectric dam owners in Maine and New Hampshire over this 21-year period document Hendel's capability to address the functionality of these vessels, from hull design to towing bitts and winches, to providing for onboard worker accommodations for 24-hour-a-day operations.

Personality and Community

Hendel's personal history of residence in bustling Hamburg, Germany, from birth to early adulthood, and then working in New York City and the Boston area frequently from 1928–31, would seem to point to living in urban areas as his personal preference. He found the coast of Maine, though, with its small towns, rural countryside, and boatyards greatly to his liking. In his résumé, he particularly notes that "most important, [he] had found many good friends," first in the Boothbay region and later in the Camden area.

Gordon Bok, son of Camden Shipbuilding Company Vice President Cary W. Bok, recalls that Hendel's office was initially on the ground floor of a small building located on the Camden Shipbuilding property and was a bright, light-filled space. He later moved across the harbor to the first floor of a building that fronted on Bay View Street but also overlooked Camden's inner harbor at Willey Wharf, owned by P.G. Willey & Co.

Bok remembered Hendel as a modest man, friendly and well-liked, who spent many Christmases at the Bok home. Bok also recalled that Hendel had three constant companions: sketch paper, a pencil or pen and ink, and a briar pipe. The aroma of pipe tobacco was a continuing accompaniment to Hendel's surroundings.

Bok recalled, too, that Hendel was an individual who applied common sense and practical solutions to problems. An example occurred in 1954 when an electric winch hauling a large vessel up Camden Shipbuilding's marine railway failed and started to unwind. Hendel took a rear wheel off the yard's work truck, replaced it with a wheel rim, and connected a long leather belt from the machine shop from the wheel rim to the winch drum. The vessel was slowly, successfully, pulled up the railway, and yard personnel joked that the truck had traveled the equivalent of 75 miles in reverse to complete the hauling-out process.

Hendel's stepdaughter, Brigitte Gefrich, described him as having a lonely childhood and thought Geerd was one who felt somewhat insecure in his earlier years. He was a good man, she noted, modest and humble, and honest and ethical.

Publisher and yacht designer Thomas MacNaughton of Pen & Anchor Publishing in Eastport, Maine (wwwmacnaughtongroup.com) writes the following about Geerd Hendel:

> Although Mr. Hendel was born in Germany, he was very much a Maine designer. He lived and worked in the Camden area for a great many years. I believe he first came to Maine in the late 1930s to work with W. Starling Burgess on the Cup Defender *Ranger*. For many years, his office overlooked Camden Harbor facing Camden Shipbuilding across the harbor where many of his designs were built.
>
> During World War II, he designed small craft for Camden Shipbuilding to construct for the Navy. After the war, he was prominent in introducing aluminum construction to boatbuilding. Later, he produced fishing draggers that were built at the Gamage yard in South Bristol, Maine.
>
> His work was varied and always interesting. It ranged from small sailboats to yachts over 100 feet long and encompassed sport fishermen and the Boothbay Harbor Yacht Club One Design.
>
> Your author remembers that Mr. Hendel was kind enough to help out a local boatbuilder, Elmer Collemer, the famous one-man boat shop. Mr. Collemer found himself in the position with one of the last boats he built of having the client wish to have Elmer take responsibility for the design as well as the construction. Elmer, though a great artist and craftsman, did not consider himself by any means a yacht designer, so he would take his drawings to Mr. Hendel for comments and any necessary calculations. When talking with Elmer about this many years ago, it immediately struck me that many designers would not have taken the time to help with a project which would not come out under their own name. Obviously Mr. Hendel's regard for the project and for Elmer made him glad to help out a fellow artist in producing a beautiful, functional work.
>
> Mr. Hendel lived to be 95 years old. Given his work, this must surely have proved a long and happy life. His entire collection of plans, photographs, and business papers relating to more than 280 designs has been donated by his wife, Hanni, to the Maine Maritime Museum. A study of these designs and their relationship to the times in which he lived would amply repay the student.
>
> Your present author remembers that he used to stand on the street staring into Mr. Hendel's office windows looking at what he could see of the pictures, the models, and the other items

> that spoke very strongly of a life spent intimately involved with design. Since I was a boy fiercely interested in yacht design, I longed to go in and meet Mr. Hendel. I never did dare to do so. I suspect that if I had ventured to go in, I might have been warmly welcomed as I would welcome a young boy or girl today. I do remember it seemed his door was always open and inviting.

For an individual who was often regarded as being shy and a loner, Hendel participated actively in the Boothbay Harbor Yacht Club, where he was on the Regatta Committee as Measurer and a member of the Entertainment Committee (!). As a long-term member of the Camden Yacht Club, he is remembered as a very humorous raconteur. Additionally, in Camden he was a decades-long member of the Camden Rotary Club, which met weekly.

Camden-based attorney Jack Sanford provided this recollection that exemplifies the quality of Geerd's character:

> In the late 1970s or early 1980s when we got our first boat, we had a mooring in Sherman's Cove [in Camden's outer harbor]. It was [and still is] a long row out there, and the rock and roll would drive you nuts, so I really wanted an inner harbor mooring. Someone suggested that I speak with Geerd, who had a very close-in harbor mooring for his power boat. Geerd was getting elderly at that point, and he was having a hard time getting from his dinghy into his boat.
>
> I suggested to Geerd that I would build a float to put on his mooring. He could easily get from his dinghy to the float, then to his boat, and I would get the other side of the float. He thought it was a great idea. I approached the harbormaster, who said we could not do it because the nearby swing moorings would interfere with a float. When I told Geerd, he said something like, "You have young kids, and you should not be rowing with them out to Sherman's Cove, so you can have my mooring [for free!]. I plan to sell my boat anyway." As you can imagine, that offer was beyond generous, as inner harbor moorings were more coveted then than inner harbor floats are now.
>
> To ease my conscience a bit, I had Neal Parent take an early-morning photo of Geerd's beloved boat in the fog. The picture was absolutely stunning. I had the picture framed, had all of our kids sign it, and presented it to Geerd as thanks. He was deeply touched, and cherished that photo until the day he died. In later years, when all the inner harbor moorings were required to add floats, I did so, and still have it. I often think of Geerd, especially when I hear that the waiting list to get an inner harbor float is now about ten years. He was a wonderful guy.

Geerd Hendel lived to be 95, and, given the quality and quantity of his design work, he surely had a fulfilling career as a naval architect. As noted in the Thomas MacNaughton commentary, his entire collection of plans, photographs, and business papers relating to more than 280 vessels was donated by his wife, Hanni, to the Maine Maritime Museum, 243 Washington Street, Bath, Maine.

The archives at the Mystic Seaport's Research Library contain the 1988 oral interview tape mentioned throughout this work. The research facility also holds the design records of W. Starling Burgess, Albert Condon, and many other designers and naval architects, along with comprehensive collections of maritime photographs, books, and yachting/boating periodicals.

APPENDIX I

Hendel Designs, Maine Maritime Museum Archives

Length	Description	Name	Client	Builder	Plan Date	Notes	MMM file location
105'	Motor yacht (Hendel developed detailed drawings for bulwark doublers and bow chocks, and changes in strakes)	*Althea*	Murphy, C.H.	BIW	1930		Box 1, MS 310
15'	One Design sloop	Maine Camp One-Design Class	Wabanaki Camp for boys, Hillside section of Sebago, Maine	Rice Brothers and Mt. Desert Yacht Yard	1931	3 built	2
16'	One Design sloop			Goudy & Stevens	1932	Robert Rice Collection	
20'	Sloop (design)		Kimberly Jr., John B		1932		MS 310
33'	Cruising ketch (design)				1932	Half-hull model built by Hendel	Box 4, MS 310
18'	Centerboard sloop				1932		MS 310
20'4"	Centerboard sloop	Boothbay Harbor 20		Goudy & Stevens	1932	*The Waterline*, BHODA	
20'4"	Centerboard sloop	Boothbay Harbor 20		Rice Brothers	1932	*The Waterline*, BHODA	
20'	Centerboard sloop	*Lao-Tse*	Attridge, The Reverend Thomas W.	Hodgdon Bros.	1932	Boothbay Region Historical Society	
55'	Steel cruiser (design)				1932	*The Rudder* magazine, August 1932	

Length	Description	Name	Client	Builder	Plan Date	Notes	MMM file location
25'	Shoal draft auxiliary sloop		Wood, Major Philip S.	Roger I. Sawyer	1934	*The Rudder* magazine, March 1934	
33'	Cruising sloop—gaff rig (design)			Goudy & Stevens	1934		4
33'	Cruising cutter (design)			Goudy & Stevens	1934	gaff rig (rolled with cruising sloop plans)	4
33'	Cruising sloop (design)			Goudy & Stevens	1934		7
23'	Auxiliary cutter			Norman Hodgdon	1934		4
26'	8-meter sloop (proposal)		Braine		1935		MS 310
33'	(design)		Designed for Hodgdon Bros.		1936	*Boothbay Register*, February 21, 1936	
24'	Auxiliary sloop	*Nema*	Sprague, Charles H.	Norman Hodgdon Jr.	1936	Design #24	4
20'4"	Keel racing sloop	*Sanderling*	Hendel, Geerd	Norman Hodgdon Jr.	1936	Design #60	8
21'	Keel sloop	*Loon*	Hendel, Geerd	Norman Hodgdon Jr.	1937	BHOD prototype	
30'	Racing sloop (Hendel as draftsman for W. Starling Burgess)	Yankee One-Design Class			1937	W. Starling Burgess, designer	4
19½'	Racing sloop (Hendel as draftsman for W. Starling Burgess)	Small Point one-design class	Small Point Yacht Club		1937	W. Starling Burgess, designer	6
15'	Aluminum dinghy		Launch to serve Thistle, a 103' bronze Herreshoff yawl	BIW	1939		MS 310, 38.106.1
21'	Racing sloop	Boothbay Harbor one-design class	Boothbay Harbor Yacht Club	Goudy & Stevens, Norman Hodgdon, and Ben Rand	1938–41	57 have been built through 2018	8
26'	Aluminum auxiliary sloop	*Whistler*	Hendel, Geerd	Rice Brothers	1939	Design #32	3

Length	Description	Name	Client	Builder	Plan Date	Notes	MMM file location
18'	Aluminum racing sloop w/keel (design)				1940	*Yachting* magazine ad	
40'	Sport fisherman/ dragger	*C.W.W.*	Named using the initials of the first owners: Carver, Whitney, and Wallace. Vessel quickly acquired by the U.S. Navy.	Camden Shipbuilding	1942	Design by Jim Stevens, with drafting assistance by Hendel	
33'	Aluminum cutter (design)				1944	1944 *Yachting* design contest, 2nd place winner	
110'	Swan's Island Ferry (proposal)		State of Maine Ferry Service	Camden Shipbuilding	1944		3
65'	Dragger	*Camden*	Knutsen, K.W.	Camden Shipbuilding	1944		5
85'	Dragger		Grinnell, Capt. Russell Jr.	Rice Brothers	1944		6
70'	Dragger (proposal)				1944		8
60'	Dragger (proposal)		Hanson, Capt. N.P.		1944		8
70'	Dragger (proposal)				1944		8
87'	Dragger (proposal)		Salvatore, Ferdinand	Gamage/ Bristol Yacht	1944		8
100'	Steam harbor tug (proposal)			Camden Shipbuilding	1945		3
9'	Aluminum rowboat				1945		6
90'	Dragger	*Wild Duck*		Gamage/ Bristol Yacht	1945		8
90'	Dragger	*The Clipper Fish*	A. Arruda	Gamage/ Bristol Yacht	1945		8
90'	Dragger	*Albatross*		Gamage/ Bristol Yacht	1945		8
61'	Dragger	*Althea Joyce*	Salvatore, Ferdinand	Gamage/ Bristol Yacht	1945		8
38'	Built as a scallop dragger, then used as a tour boat	*Betselma*	Wooster, Ralph E.	Camden Shipbuilding	1945		5

Length	Description	Name	Client	Builder	Plan Date	Notes	MMM file location
33'	Sloop (proposal)		Peterson, William "Pete"		1946	Camden Public Library	
40'	Sport fisherman	*Andrea B*	Boardman, Earl	Camden Shipbuilding	1946	Malcolm Brewer design with Hendel drafting the plans	2
40'	Sport fisherman	*Marbaru*	Lunt, Clinton	Camden Shipbuilding	1946	Malcolm Brewer design with Hendel drafting the plans	2
16'	Aluminum centerboard sloop			Camden Shipbuilding	1946	Production boat	2
18'	Keel racing sloop, aluminum			Camden Shipbuilding	1946-48	Production boat	2
28'	Power cruiser (design)			Camden Shipbuilding	1946		2
42'	Sport fisherman (variations)			Camden Shipbuilding	1946		5
37'	Aluminum auxiliary sloop (design)				1946	*Yachting* ad, January 1946	MS 310
84'	Auxiliary ketch				1946		5
18'	V-bottom aluminum canoe				1946		6
40'	Utility launch		Camden Shipbuilding	Camden Shipbuilding	1946		6
40'	Power cruiser		Deering, A. Clement	Camden Shipbuilding	1946	Malcolm Brewer design with Hendel drafting the plans	6
99'	Dragger	*Pam-Ann*	Grinnell, Capt. Russell	Casey Boat Building Co.	1946		8
32'	Auxiliary sloop	*Godofin*	Chatfield, Albert H.	Camden Shipbuilding	1946		8
36'	Power cruiser/ Lobster boat	*Madelyn E*	Young, Bradbury	Camden Shipbuilding	1947		4
64'	Sardine carrier (proposal)			Camden Shipbuilding	1947	60' and 64' options for same vessel	6

Length	Description	Name	Client	Builder	Plan Date	Notes	MMM file location
60'	Sardine carrier (proposal)			Camden Shipbuilding	1947	60' and 64' options for same vessel	6
16'	Aluminum centerboard cat boat			Camden Shipbuilding	1947		6
16'	Centerboard sloop, aluminum			Camden Shipbuilding	1947		7
16' & 18'	Aluminum sloops—mast arrangement				1947		6
20'	Centerboard sloop			Kennebec Crafts	1947	Design #75	6
18'	Keel cat boat—aluminum			Camden Shipbuilding	1947		6
22'	Sport cruiser/bass boat			Camden Shipbuilding	1947	Production boat	6
42'	Auxiliary double-headsail sloop	*Alphard*	Bok, Curtis	Camden Shipbuilding	1947	Re-rigged as Ketch 1958	7
32'	Diesel aluminum towboat (proposal)		Brown Paper Co.	Camden Shipbuilding	1947		9
36'	"Maine Coast 36" sloop (design)	3 were built; *Snowflake*, *Torna*, and *Suliko*			1947	Hendel as consultant to E. Farnum Butler, N.A.	MS 310
42'	Auxiliary sloop	*Gone Away*	Cheston, Morris	Camden Shipbuilding	1947		3
42'	Auxiliary sloop	*Gone Away*	Cheston, Morris	Camden Shipbuilding	1947		10
55'	Cruiser	*Northern Cross*	Bok, Cary	Camden Shipbuilding	1947	Camden Public Library	
32'	Power boat			Goudy & Stevens	1948		6
14'	V-bottom centerboard sloop			Goudy & Stevens	1948		6
44'	Sardine carrier	*Alice M*	Ross, Emery S.	Goudy & Stevens	1948		7
40'	Commercial sport fisherman		Marsh, Leslie	Camden Shipbuilding	1948		7
55'	Diesel power cruiser	*Fantasy*	Parson, Donald	Camden Shipbuilding	1948		10

Length	Description	Name	Client	Builder	Plan Date	Notes	MMM file location
98'	Passenger boat, twin-screw				1948		10
69'	Motor sailer/ research vessel, double-ended				1948	*The Rudder*, April 1948	MS 310
35'	Power boat	*Snapper*	Saltonstall, Leverett	Camden Shipbuilding	1949	Drafting for Malcolm Brewer's design	3
21'	Utility launch		Sonia, Harold J.		1949		4
38'	Twin-screw sport fisherman (design)			Goudy & Stevens	1949		5
28'	Sport fishing launch		Burnside, Irving H.	Camden Shipbuilding	1949		5
46'	Seine boat			Goudy & Stevens	1949		6
28'	Sport cruiser		Barker, B. Devereux	Camden Shipbuilding	1949		6
48'	Twin screw diesel cruiser			Camden Shipbuilding	1949		6
42'	Twin-screw sport fishing cruiser (design)		Aspinwall, J.C.	Camden Shipbuilding	1949		6
40'	Log-boom tug	*Bigtow*	Central Maine Power Co.	General Foods Corp.	1949		MS 310
33'	Auxiliary racing sloop (design)			Camden Shipbuilding	1949		7
28'	Sport fishing launch		Burnside, Irving H.	Camden Shipbuilding	1949		7
22'	Bass boat, aluminum	*Chausser*	Gardiner	General Seafoods Corp.	1949	Camden Public Library	
40'	Twin-screw cruiser (proposal)		Darling, Ira		1950	Darling purchased a 40' Chris-Craft instead	2
40'	Auxiliary ketch		Phinney, Charles W.	Camden Shipbuilding	1950	Production boat	3
38'	Auxiliary sloop	*Andiamo*	Wipple, William K.	Elmer Collemer	1950		4

Length	Description	Name	Client	Builder	Plan Date	Notes	MMM file location
42'	Twin-screw sport fisherman	*Wild Goose*	Built for Peters, W.H., but sold to Eberstadt, Ferdinand while under construction	Camden Shipbuilding	1950		6
35'	Tow and winch boat	*GNP Co. No. 45*	Great Northern Paper Co.	General Seafoods Corp.	1950		9
35'	Tow and winch boat	*GNP Co. No. 46*	Great Northern Paper Co.	General Seafoods Corp.	1950		9
35'	Power boat		Bok, Cary		1950		10
70'	Diesel cruiser	*Anihita IV*	Epstein, Alfred	Camden Shipbuilding	1950		10
64'	Twin-screw passenger boat	*Sunbeam*	Deer Island Granite Co.		1951		MS 310
61'	Swordfisherman	*Porpoise*	Laytham, Allison	Camden Shipbuilding	1951		1
61'	Twin-screw diesel cruiser	*Aweigh*	Requardt, Fred		1951		2
61'	Twin-screw diesel cruiser	*Brasil*	Wiseman, John M.		1979	Rebuild of *Aweigh*	1
50'	Shrimp trawler (proposal)		General Foods Corp.	Camden Shipbuilding	1951		6
28'	Log-boom tug	*Middletow*	Central Maine Power Co.	General Foods Corp.	1951		MS 310
22'	Open boat/striped bass fisherman	*Sea Coot*	Hall, Winslow	Camden Shipbuilding	1952	Camden Public Library	
22'	Open boat	*Night Owler*	Owler, Donald	Camden Shipbuilding	1952	Camden Public Library	
22'	Open boat, aluminum	*Russell B*		Camden Shipbuilding	1952	Camden Public Library	
40'	Auxiliary sloop				1952		2
68'	Diesel yacht	*Sooner*	Davis, E. Julian	Camden Shipbuilding	1952		3
107'	Training ship (proposal)		Isbrandtsen Co.	Camden Shipbuilding	1952		4
40'	Auxiliary sloop			Goudy & Stevens	1952		4
36'	Utility boat conversion		Brown Paper Co.	General Seafoods Corp.	ca. 1952		6

Length	Description	Name	Client	Builder	Plan Date	Notes	MMM file location
77'	Diesel screw fishing vessel	*Ruth & Moses*		Harvey Gamage/ Bristol Yacht	1952	Admeasurement plan	5
43'	Diesel tug	*Nibroc*	Brown Paper Co.	General Seafoods Corp.	1952		6
92'	Dragger	*Pocahontas*	Pocahontas Fishing Co.	Snow Shipyard	1952	proposed as 87'	8
24'	Steel tug	*Kennebec No. 3*	Kennebec Log Driving Co./ Central Maine Power Co.	General Seafoods Corp.	1952		8
27'	Steel tug	*Papoose*	Kennebec Log Driving Co./ CMP	General Seafoods Corp.	1952		8
19'	Steel workboat		Great Northern Paper Co.	General Seafoods Corp.	1952		9
26'	Auxiliary sloop (design)		Sawyer, Jonathan		1953		7
27'	Diesel tug	*Little Squaw*	Kennebec Log Driving Co./ Central Maine Power Co.	General Seafoods Corp.	1953		8
33'	River tank barge		Brown Paper Co.	General Seafoods Corp.	1953		8
44'	Steel diesel tug	*Big Squaw*	Kennebec Log Driving Co./ CMP	General Seafoods Corp.	1953	For use on Indian Pond	9
40'	Sport cruiser (design)		Greene, S.		1953		MS 310
17'	Racing sloop—sail plan	Dark Harbor 17	Peterson, William "Pete"	Bohndell Sails	1954		7
35'	Steel towboat	*GNP Co. No. 49*	Great Northern Paper Co.	General Seafoods Corp.	1954		9
35'	Steel towboat	*Whitecap*	Hollingsworth & Whitney Paper Co.	General Seafoods Corp.	1954	Bert Snow, The Main Beam	
22'	Aluminum bass boat		ALCOA	General Seafoods Corp.	1954		2
23'	Harbor tug	*Barbie D*	Camden Shipbuilding	Camden Shipbuilding	ca. 1954		9
98'	Diesel cruiser (proposal)	*Vedersein*	Laytham, Allison		1955		3

Length	Description	Name	Client	Builder	Plan Date	Notes	MMM file location
110'	Super-yacht	*Vedersein*	Laytham, Allison	Abeking & Rasmussen	1955		MS 310
	Davit, chocks, & bollard	*Vedersein*	Laytham, Allison	Camden Shipbuilding	1957		7
47'	Twin-screw cruiser		Nevins, M.J.	Camden Shipbuilding	1955		5
63'	Diesel cruiser		Payson, Charles S.		1955		5
26'	Auxiliary sloop	*Widgeon*, later *Jackie B*	Foot, S.B.	Norman Hodgdon	1955-57		6
27'	Steel diesel tug	*Whitecap*	Hollingsworth & Whitney Co.	General Seafoods Corp.	1955		8
43'	Gas fishing and passenger boat	*Bingo IV*	Lovgren, Knute	Axel Gronros's Rockland Boat Shop	1956		3
73'	Diesel yacht	*Admiral Blake*	Blair, John	Camden Shipbuilding	1956		
18'	Workboat		Kennebec Log Driving Association	Goudy & Stevens/ Hodgdon Bros. Partnership	1956	Goudy & Stevens hull list	5
18'	Workboat		Kennebec Log Driving Association	Goudy & Stevens/ Hodgdon Bros. Partnership	1956	Goudy & Stevens hull list	
20'	Workboat		St. Croix Paper Co.	Goudy & Stevens/ Hodgdon Bros. Partnership	1956	Goudy & Stevens hull list	
	Steam tug boiler installation	*Walter Ross*	Ross Towing Co.		1956		6
22'	Cutter		Green, Lucien		1956		6
42'	Diesel cruiser (proposal)		Simonds, Daniel	Camden Shipbuilding	1957		3
32'	Twin-screw cruiser (design)				1957		6
60'	Twin-screw cruiser (design)		Windle		1957		6
45'	Twin-screw cruiser		Windle	Camden Shipbuilding			6
68'	Auxiliary ketch (design)		Anderson, E. Ross		1957		6

Length	Description	Name	Client	Builder	Plan Date	Notes	MMM file location
35'	Log-boom jumper	*GNP Co. No.* ___	Great Northern Paper Co.	Goudy & Stevens/ Hodgdon Bros. Partnership	1957	Fogler Library, Univ. of Maine, Orono	
35'	Log-boom jumper	*GNP Co. No.* ___	Great Northern Paper Co.	Goudy & Stevens/ Hodgdon Bros. Partnership	1957		2
35'	Log-boom jumper	*GNP Co. No. 57*	Great Northern Paper Co.	Goudy & Stevens/ Hodgdon Bros. Partnership	1957	Fogler Library, Univ. of Maine, Orono	
28'	Workboat	*GNP Co. No.* ___	Great Northern Paper Co.	Goudy & Stevens/ Hodgdon			MS 310
42'	1947 sloop re-rigged as ketch	*Alphard*	Bok, Curtis	Camden Shipbuilding	1957		7
40'	Ketch	*The Frances* or *Frances*	Merryman, Philip E.	Camden Shipbuilding	1962	Production boat	3
70'	Cruiser	*Pearl Necklace*	Maine Development Corp.	Camden Shipbuilding	1958		1
14'	Aluminum outboard motorboat/launch	*Pearl Necklace*	Maine Development Corp.	Goudy & Stevens	1958		1
10'	Aluminum pram	*Pearl Necklace*	Maine Development Corp.		1958		1
8'	Plexiglass-bottom pram	*Pearl Necklace*	Maine Development Corp.		1958		1
33'	Sportster cruiser	*Shada*	Sortwell, Daniel R. Sr.	Camden Shipbuilding	1958	w/captain's chair	6
33'	Sportster cruiser	*Tally Ho*	Chatfield, William	Camden Shipbuilding	1958		MS 310
38'	Motor sailer				1958		3
33'	Twin-screw sport fisherman/cruiser	*Project*	Brown, William (Mr. and Mrs.)	Camden Shipbuilding	1958	Camden Public Library	
14'	Aluminum outboard motorboat		Central Maine Power Co.	Goudy & Stevens	1958		4

Length	Description	Name	Client	Builder	Plan Date	Notes	MMM file location
18'	Aluminum punt		Central Maine Power Co.		ca. 1958		8
26'	Auxiliary sloop	*Jackie B* (formerly *Widgeon*)	Bates, Murray	Refit at Sid Norton's yard	1959		1
63'	Tug (proposal)		Great Northern Paper Co.		1958		9
35'	Auxiliary (diesel) sloop		Green		1959		7
	Boatel Marina, Boothbay Harbor		Neily Rupert Jr., Maine Coast Boat Sales		1959		4
53'	Cruiser	*La Reine*	Watson, Carlisle V.	Newbert & Wallace	1959		6
35'	Diesel sloop (proposal)				1959		7
63'	Twin-screw cruiser (design)		Dixon, Arthur		1960		1
42'	Ketch (design)	Robinhood 42		Kennebec Yacht Building Corp.	1960	None were built	2
33'	Twin-screw sport cruiser/lobster/ pleasure	*The Rockhopper*	Dewees, Dr. David C.	Newbert & Wallace	1960	Rolled with *Alster* plans	2
33'	Sportster cruiser	*Sabrina or Sabina*	Jones, Paul, M.D.	Camden Shipbuilding	1960	Camden Public Library	
40'	Twin-screw cruiser				1960		10
44'	Twin-screw sport cruiser				1960 & 1966		1
64'	Staysail schooner (design)		Bok		1961	See also 1981	5
82'	Diesel Cruiser (design)		Laytham, Allison		1961		MS 310
32'	Cutter	*Tahra*		Bigelow Bros.	1961		Wooden Boat Registry
70'	Twin-screw steel tug	*William Hilton*	Great Northern Paper Co.	Goudy & Stevens	1961		9
40'	Diesel flying-bridge cruiser	*Windless*	Winslow, Anthony W.	Camden Shipbuilding	1961		1

Length	**Description**	**Name**	**Client**	**Builder**	**Plan Date**	**Notes**	**MMM file location**
21'	Sloop, racing (design)	Hodgdon 21			1962	Plastic hull	3
36'	Patrol boat (proposal)		State of CT Board of Fisheries & Game		1963		2
36'	Ketch				1963		2
65'	Motor vessel (proposal)	*Sunbeam IV*	Maine Seacoast Mission Society		1963		MS 310
43'	Steel pilot vessel (proposal)		Penobscot Bay and River Pilots Association		1963		2
30'	Auxiliary sloop				1963		2
49'	Auxiliary ketch				1963		2
25'	Friendship sloop	*Old Baldy*	Roy, Charles	Bald Mountain Boat Works	1963	Alterations to finish a 1971 Jarvis Newman hull based on 1914 *Pemaquid*	4
30'	Log-boom tender	*GNP Co. No. __*	Great Northern Paper Co.	Goudy & Stevens	1963	Goudy & Stevens hull list	
36'	Steel tug (proposal)		Zahn, Bernard T.		1963		
30'	Auxiliary Friendship sloop	*Heritage*	Heritage, William, and Ellsworth, Sandy	Elmer Collemer builder, Murray Peterson designer, Geerd Hendel design consultant	1963		5
31'	Sport fisherman				1963		6
70'	Twin-screw steel tug	*O.A. Harkness*	Great Northern Paper Co.	Goudy & Stevens	1964		8
50'	Twin-screw cruiser with fishing cockpit (design)				1964	3 variations of the same vessel	9
60'	Twin-screw cruiser (design)				1964	3 variations of the same vessel	1

Length	Description	Name	Client	Builder	Plan Date	Notes	MMM file location
	Auxiliary sloop	*Mooncusser*			1964		1
34'	Lobster boat			Reinforced Plastics Co.	1964		2
27'	Lobster boat	*Wandel*	Kunkle, Dr. E.C.	Sam Guild Boat Shop	1964		MS 310
27'	Research boat (lobster boat hull)		Michigan State University	A.T. Greene, boatbuilder	1965	Identical to *Wandel*	1
49'	Ketch (design)		Clapp, N.D.		1965		1
	Diesel cruiser		Hulbert, W.R.		1965	Refrigerator	MS 310
46'	Twin-screw sport cruiser (proposal)		M.Y.A.B.C.A. (Mallorca Yacht and Boat Construction Association)	Mallorca Yacht	1965		1
25'	Friendship sloop	*Sazerac*	McKinnon, Georges	Bald Mountain Boat Works	1965	Lines taken from Howard Chapelle's 1935 drawings for *Pemaquid*	1
27'	PADEBCO cruiser	*Brennain*	Neeson, Jack and Margaret	PADEBCO Custom Boats	1965	Cruiser customized	2 MS 310
36'	Cruiser (interior layout)		Bald Mountain Boat Works	Bald Mountain Boat Works	ca. 1965	Likely a duplicate of 36' boat built in 1968	
40'	Sport fisherman, Downeast 40 production model hull				ca. 1965	Double cabin layout	1
38'	Steel tug/workboat	*Seasmoke*	Post Lobster Pound, Tenants Harbor	Thomaston Steel Works	1965		1
101'	Schooner yacht	*America* (replica)	Schaefer, Rudolph Jr.	Goudy & Stevens	1966	Tank capacity curve detail	5
70'	Wooden schooner pilot boat				1966		1
40'	Dragger (proposal)				1966		2
35'	Auxiliary cutter, double headsail	*Lady Joan*	Greene, Col. A.T.	Morse Boatbuilding	1966	Camden Public Library	2

Length	Description	Name	Client	Builder	Plan Date	Notes	MMM file location
62'	Pilot vessel (proposal)		Portland Pilots, Inc. (Maine)		1966		
22'	Aluminum lobster boat, stern drive		Post Lobster Pound, Tenants Harbor	Thomaston Steel Works	1966		2
20'	Steel workboat		Bickford, Clyde L.		1966		5
27'	Lobster boat/day cruiser	*Vicil*	Sargent, Philip A.	A.T. Greene, boatbuilder	1966		5
35'	Twin-screw fisherman cruiser with flying bridge				1967		MS 310
40'	Double-ended dragger-type cruiser (design)		Main, B.A. Jr.		1967		1
40'	Twin-screw diesel cruiser		Sargent	Morse Boatbuilding	1967		1
40'	Customized cabin for a Dyercraft hull	*Blue Pigeon*	Krementz, Richard	Camden Shipbuilding	1971		1
45'	Offshore cruiser, double-ended, "Bald Mountain 45"	*Kyloe*	Noyes, Edward R. Jr.	Bald Mountain Boat Works	1967		1
45'	Double-ended cruiser (*Kyloe*?)				1967		2
65'	Cruiser yacht, twin-screw diesel (design)		Hoffberger, Charles C. II		1967		1
40'	Ketch, production boat	*Freya*	Parson, Donald	Lee's Boat Shop	1967	Camden Public Library	2
34'	Auxiliary cruising sloop				1967		
	Twin-screw diesel yacht				1967		2
69'	Double-ended twin-screw diesel yacht		Quimby		1967		1
42'	Power cruiser "Explorer" (design)				1968		2
38'	Auxiliary ketch				1968 and 1974	With or without deckhouse	1
38'	Passenger fishing vessel, diesel				1968		2

Length	Description	Name	Client	Builder	Plan Date	Notes	MMM file location
36'	Lobster-style boat (Ken Smith design)	*Pebbles*	Rockefeller, James Jr.	Bald Mountain Boat Works	1968	Interior layout	2
27'	Lobster boat	*Wandel*	Kunkle. Dr. E.C.	Sam Guild Boat Shop	1968		2
23'	Auxiliary cutter			Norman Hodgdon	1968		1
	Motel, home and restaurant in Boothbay (plan)		Wendland, Mr. & Mrs. Max		1968	4-unit motel, 19-seat restaurant, and residence in one structure	2
	Aluminum submarine tender (2 versions)				1968		3
	Sea moss catamaran		Emerson, E.W.		1968		5
30'	Open lobster boat	*Evie Sue*	Ames, Crosby	Gerry Libby's Melvin Heights Boat Shop	1968		5
33'	Schooner	*Felicity*	Manigault, Pierre	Elmer Collemer and Southwest Boat	1968	Designed by Hendel and Elmer Collemer	MS 310
	Passenger and freight catamaran				1968		
36'	Lobster boat-style hull		Brierly, Ernest M.		1969		5
38'	Pilot and lobster boat	*Phalarope*	Stanley, Sherman	Gerry Libby's Melvin Heights Boat Shop	1969		1
33'	Express cruiser (Sportster)	*Alster*	Hendel, Geerd	Old Kiln Boat Works, Inc.	1969		2
124'	Twin-screw yacht	*Claybeth*	Gengras, E. Clayton	Abeking & Rasmussen	1969		2
30'	Sloop				1971		8
40'	Aluminum fishing vessel				1971		1
31'	Ketch sail & rigging plan for a Concordia 31			Wayfarer Marine	1971		1

Length	Description	Name	Client	Builder	Plan Date	Notes	MMM file location
32'	Friendship sloop, double headsail	*Diana*	Gay, Ebenezer and Diana	Hull built by Jarvis Newman Yard. Finished by Bald Mountain Boat Works	1971	Hendel-designed interior layout	2
40'	Diesel cruiser (proposal)				1972		Original owner
70'	Diesel yacht	*Moby Dick IV*	Thompson, Peter		1972	Design for after deck enclosure and new flying bridge	1
48'	Lee cutter lines, plan, and section		Highland Community School, Scroon Lake, Paradox, NY	Based on an H. I. Chapelle plan	1972	2	1
44'	Glass-bottom boat (proposal)		Williams Marine, Ltd. Hamilton, Bermuda			1972	
54'	Auxiliary aluminum yawl, new arrangement	*Doric*	Tanenbaum, Stanley	Abeking & Rasmussen	1972	W.H. Tripp, original designer (1966)	5
38'	Auxiliary ketch (design)		Chatterjee, Dr. Manu		1973		7
	Patrol and research vessel refit of 1964 GNP Co. logging tug *O.A. Harkness*	*O.A. Harkness*	Maine Dept. of Sea and Shore Fisheries		1973		MS 310
70'							
70'	Twin-screw steel tug refit of 1960 GNP Co. logging tug *William Hilton*	*Argonaut*	Winslow, Eliot, of Winslow Marine, Inc.		1973		2
40'	50-passenger boat with Down East 40 hull		Thurston Co., Orland, Maine		970s	Thurston Co. finished Down East 34 and 40 hulls	5
61'	Steel-hull diesel cruiser	*Aweigh*	Requardt, J. Frederic Jr.		1974	Client built in wood and altered transom	PL-19 Box 1

Length	Description	Name	Client	Builder	Plan Date	Notes	MMM file location
64'	Tug	*Alton A. II*	Prock Marine, Rockland, Maine	Damariscotta Boats, Inc.	1975	Built as 44'	8
45'	Two-masted sailing vessel				1980		3
40'	Fishing boat with auxiliary sail and fwd pilothouse				1980		1
24'	Powered yawl boat				1982		1
18'	Aluminum sloop				1984	New keel	4
15'	Aluminum skiff				1986		2
60'	Sport fisherman				1986		4
64'	Coastal voyager with two staysail masts (design)		Bok		1981	See also 1961	10
33'	Power boat (Sportster design)		Williams, John		1989	Based on 1958 design for Bill Brown	1
	Trimaran aluminum wishbone boom details, and fittings	*Caha Paluna II*		Wayfarer Marine			3
26'	Harbormaster workboat, aluminum	*Welcome*	Town of Camden	Paul E. Luke Yard	1989		8
33'	Launch		Fish & Wildlife Service (unknown state)		1990		4
40'	Construction barge				1990		4
75'							5
130'	Ferry (proposal)		Maine Port Authority			Penobscot	3
64'	Steel passenger boat						3
40'	Diesel cruiser		Dabeney, Thomas				4
65'	Motor tug—rudder only		General Seafoods	General Seafoods			5

APPENDIX II

Artifacts and Professional Documents

Geerd and sister Marianne at the Isle of Juist, 1904–05.
Barthold and Ingrid Brümmer Collection.

Geerd and sister Marianne at the Isle of Juist 1907–08.
Barthold and Ingrid Brümmer Collection.

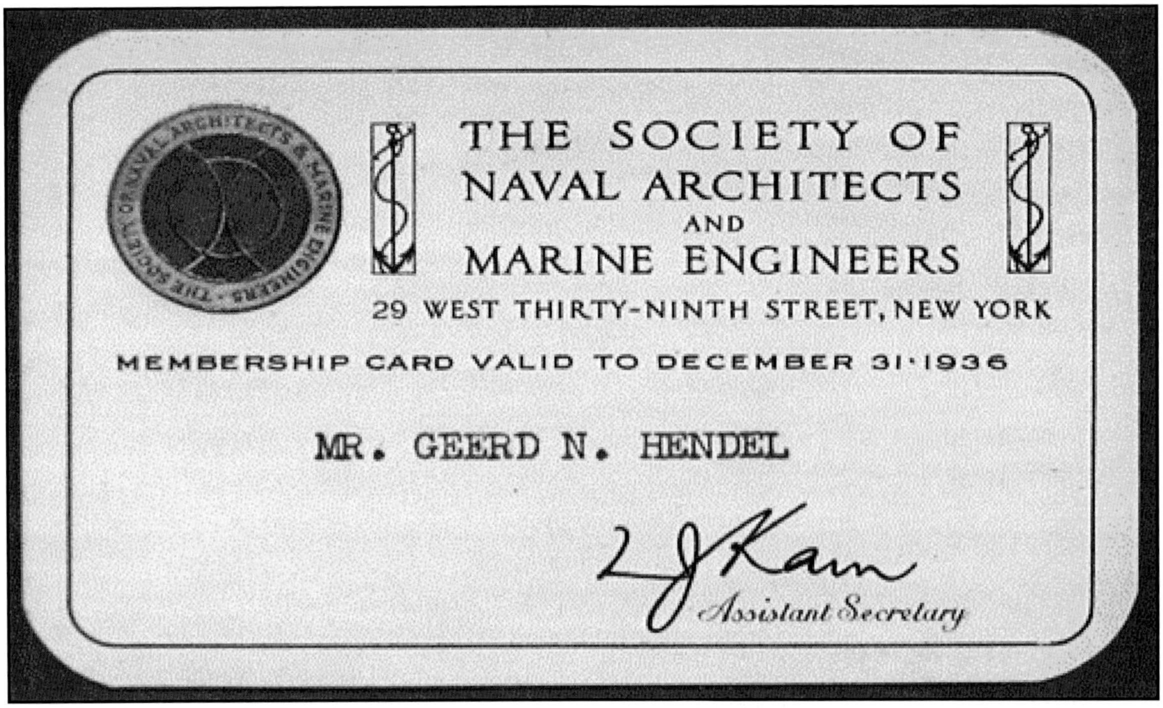

THE SOCIETY OF NAVAL ARCHITECTS AND MARINE ENGINEERS

29 WEST THIRTY-NINTH STREET, NEW YORK

MEMBERSHIP CARD VALID TO DECEMBER 31·1936

MR. GEERD N. HENDEL

Assistant Secretary

Hendel became a member of the Society of Naval Architects and Marine Engineers in 1929, and his 1936 membership card is included in archival files.

Hendel Photograph Collection, Box MS-310, Maine Maritime Museum.

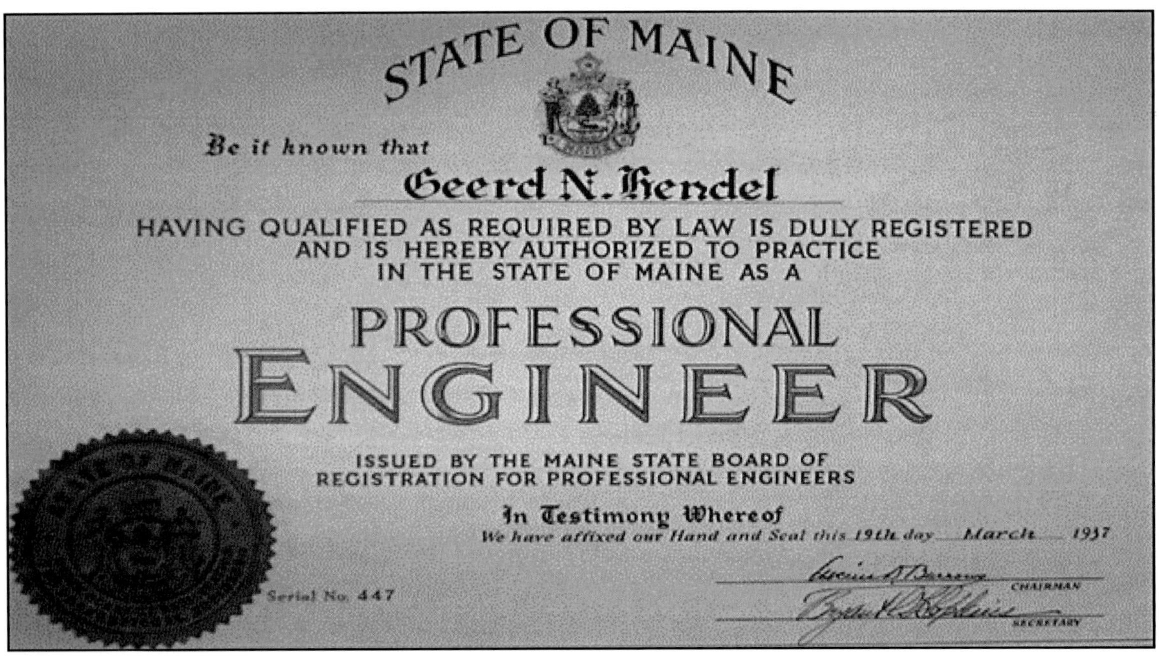

STATE OF MAINE

Be it known that

Geerd N. Hendel

HAVING QUALIFIED AS REQUIRED BY LAW IS DULY REGISTERED AND IS HEREBY AUTHORIZED TO PRACTICE IN THE STATE OF MAINE AS A

PROFESSIONAL ENGINEER

ISSUED BY THE MAINE STATE BOARD OF REGISTRATION FOR PROFESSIONAL ENGINEERS

In Testimony Whereof

We have affixed our Hand and Seal this 19th day March 1937

Serial No. 447

CHAIRMAN

SECRETARY

Hendel's State of Maine Professional Engineer's License issued March 19, 1937.

Hendel Photograph Collection, Box MS-310, Maine Maritime Museum.

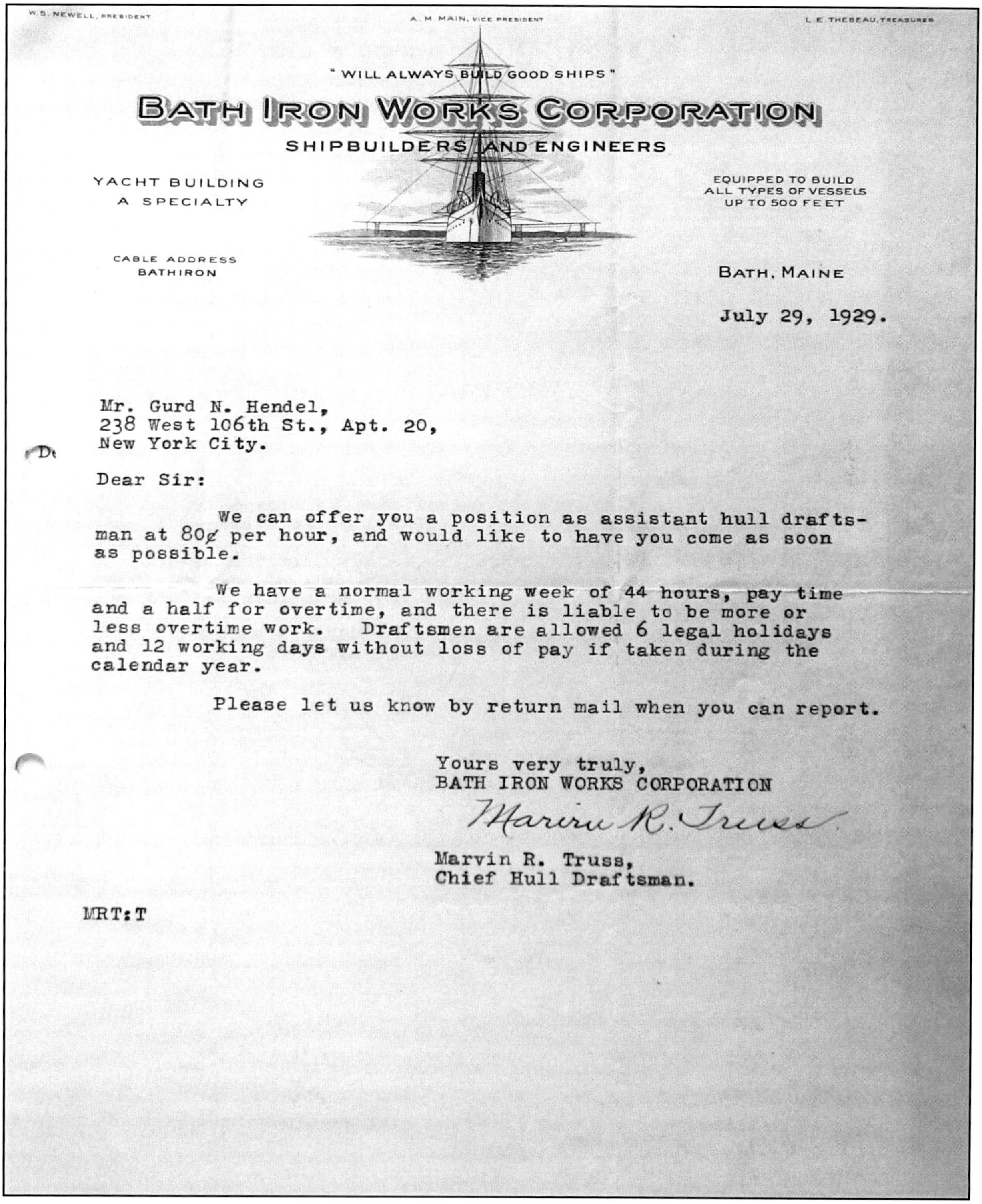

W. S. NEWELL, PRESIDENT A. M. MAIN, VICE PRESIDENT L. E. THEBEAU, TREASURER

" WILL ALWAYS BUILD GOOD SHIPS "

BATH IRON WORKS CORPORATION

SHIPBUILDERS AND ENGINEERS

YACHT BUILDING
A SPECIALTY

EQUIPPED TO BUILD
ALL TYPES OF VESSELS
UP TO 500 FEET

CABLE ADDRESS
BATHIRON

BATH, MAINE

July 29, 1929.

Mr. Gurd N. Hendel,
238 West 106th St., Apt. 20,
New York City.

Dear Sir:

We can offer you a position as assistant hull draftsman at 80¢ per hour, and would like to have you come as soon as possible.

We have a normal working week of 44 hours, pay time and a half for overtime, and there is liable to be more or less overtime work. Draftsmen are allowed 6 legal holidays and 12 working days without loss of pay if taken during the calendar year.

Please let us know by return mail when you can report.

Yours very truly,
BATH IRON WORKS CORPORATION
Marvin R. Truss
Marvin R. Truss,
Chief Hull Draftsman.

MRT:T

Hendel's July 29, 1929, letter of employment at BIW.

Hendel Photograph Collection, Box MS-310, Maine Maritime Museum.

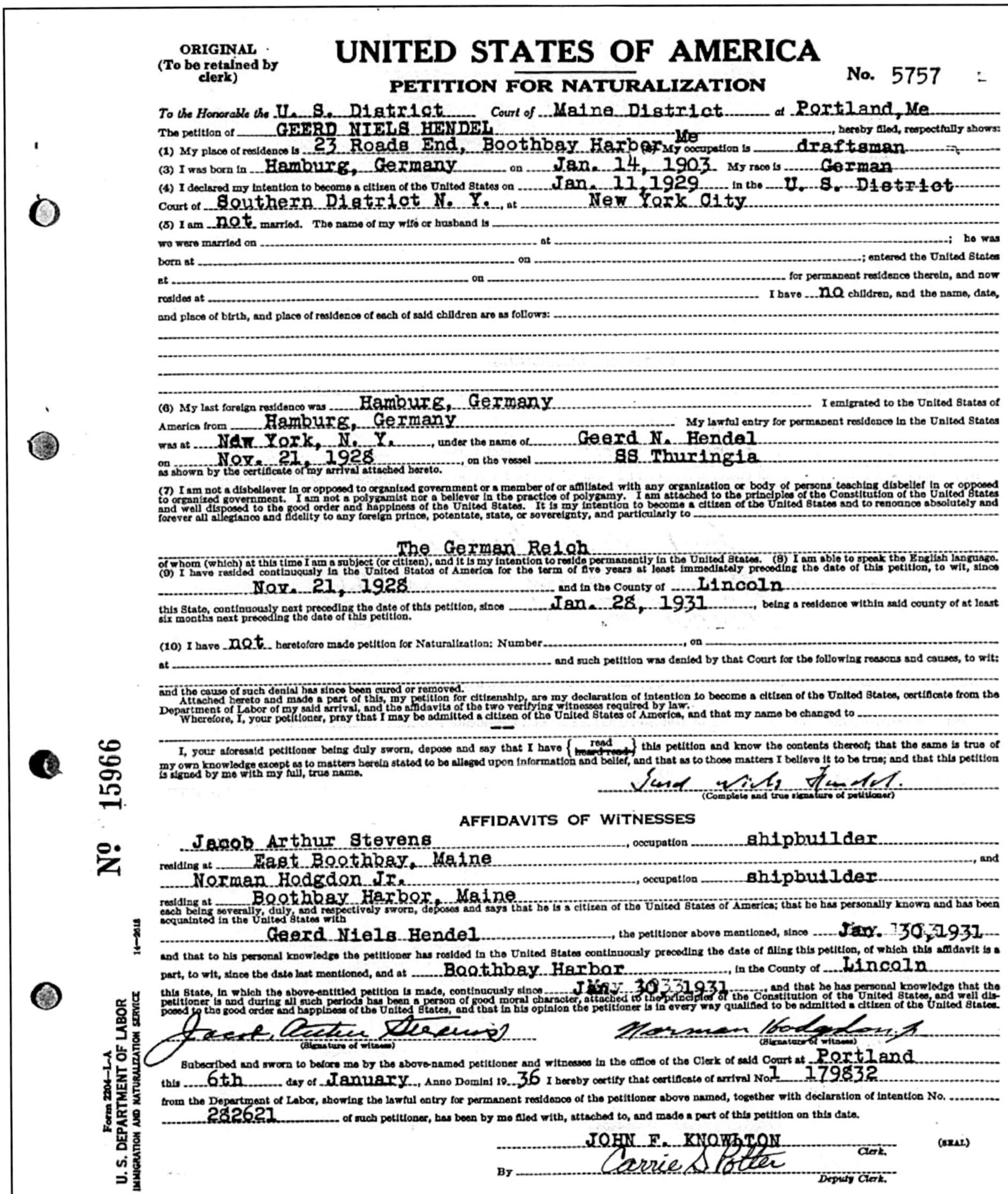

ORIGINAL (To be retained by clerk)

UNITED STATES OF AMERICA

PETITION FOR NATURALIZATION

No. 5757

To the Honorable the U. S. District Court of Maine District at Portland, Me

The petition of GEERD NIELS HENDEL, hereby filed, respectfully shows:

(1) My place of residence is 23 Roads End, Boothbay Harbor, Me (2) My occupation is draftsman

(3) I was born in Hamburg, Germany on Jan. 14, 1903 My race is German

(4) I declared my intention to become a citizen of the United States on Jan. 11, 1929 in the U. S. District Court of Southern District N. Y. at New York City

(5) I am not married. The name of my wife or husband is .... we were married on .... at ....; he was born at .... on ....; entered the United States at .... on .... for permanent residence therein, and now resides at .... I have no children, and the name, date, and place of birth, and place of residence of each of said children are as follows: ....

(6) My last foreign residence was Hamburg, Germany I emigrated to the United States of America from Hamburg, Germany My lawful entry for permanent residence in the United States was at New York, N. Y., under the name of Geerd N. Hendel on Nov. 21, 1928, on the vessel SS Thuringia as shown by the certificate of my arrival attached hereto.

(7) I am not a disbeliever in or opposed to organized government or a member of or affiliated with any organization or body of persons teaching disbelief in or opposed to organized government. I am not a polygamist nor a believer in the practice of polygamy. I am attached to the principles of the Constitution of the United States and well disposed to the good order and happiness of the United States. It is my intention to become a citizen of the United States and to renounce absolutely and forever all allegiance and fidelity to any foreign prince, potentate, state, or sovereignty, and particularly to The German Reich of whom (which) at this time I am a subject (or citizen), and it is my intention to reside permanently in the United States. (8) I am able to speak the English language. (9) I have resided continuously in the United States of America for the term of five years at least immediately preceding the date of this petition, to wit, since Nov. 21, 1928 and in the County of Lincoln this State, continuously next preceding the date of this petition, since Jan. 28, 1931, being a residence within said county of at least six months next preceding the date of this petition.

(10) I have not heretofore made petition for Naturalization: Number ...., on .... at .... and such petition was denied by that Court for the following reasons and causes, to wit: .... and the cause of such denial has since been cured or removed.

Attached hereto and made a part of this, my petition for citizenship, are my declaration of intention to become a citizen of the United States, certificate from the Department of Labor of my said arrival, and the affidavits of the two verifying witnesses required by law.

Wherefore, I, your petitioner, pray that I may be admitted a citizen of the United States of America, and that my name be changed to ....

I, your aforesaid petitioner being duly sworn, depose and say that I have read this petition and know the contents thereof; that the same is true of my own knowledge except as to matters herein stated to be alleged upon information and belief, and that as to those matters I believe it to be true; and that this petition is signed by me with my full, true name.

Geerd Niels Hendel (Complete and true signature of petitioner)

AFFIDAVITS OF WITNESSES

Jacob Arthur Stevens, occupation shipbuilder residing at East Boothbay, Maine, and Norman Hodgdon Jr., occupation shipbuilder residing at Boothbay Harbor, Maine each being severally, duly, and respectively sworn, deposes and says that he is a citizen of the United States of America; that he has personally known and has been acquainted in the United States with Geerd Niels Hendel, the petitioner above mentioned, since Jan. 30, 1931 and that to his personal knowledge the petitioner has resided in the United States continuously preceding the date of filing this petition, of which this affidavit is a part, to wit, since the date last mentioned, and at Boothbay Harbor, in the County of Lincoln this State, in which the above-entitled petition is made, continuously since Jan. 30, 1931, and that he has personal knowledge that the petitioner is and during all such periods has been a person of good moral character, attached to the principles of the Constitution of the United States, and well disposed to the good order and happiness of the United States, and that in his opinion the petitioner is in every way qualified to be admitted a citizen of the United States.

Jacob Arthur Stevens (Signature of witness) Norman Hodgdon Jr. (Signature of witness)

Subscribed and sworn to before me by the above-named petitioner and witnesses in the office of the Clerk of said Court at Portland this 6th day of January, Anno Domini 1936 I hereby certify that certificate of arrival No. 1 179832 from the Department of Labor, showing the lawful entry for permanent residence of the petitioner above named, together with declaration of intention No. 282621 of such petitioner, has been by me filed with, attached to, and made a part of this petition on this date.

JOHN F. KNOWLTON Clerk. (SEAL)

By Carrie S Potter Deputy Clerk.

No. 15966

Form 2204—L-A U. S. DEPARTMENT OF LABOR IMMIGRATION AND NATURALIZATION SERVICE 14—2613

Hendel's petition for naturalization as a citizen of the United States.

Terri Reynolds Collection (Norman Hodgdon Jr.'s granddaughter).

P.O. BOX 337
CAMDEN, MAINE
04843

GEERD N. HENDEL
NAVAL ARCHITECT

AREA CODE 207
236-2686

(Resume) January 1989.

Geerd Niels Hendel. (Husband of Hanni Hendel).
Born: 1 - 14 - 1903. in Hamburg, Germany.
Parents: Justus and Clara (Girardet) Hendel.

Education: Public schools in Hamburg, then, after 2 years of shipyard (Deutsche Werft, Hamburg) apprenticeship, attended the Higher Technical Institute (5 semesters) at Bremen, Germany, specialising in Naval Architecture and graduating in 1925.

Employment: Two years plus in the design office of a very active shipyard with a crew of about 250 men building small commercial vessels of steel, Werft Nobiskrug, Rendsburg in Northern Germany.

Following: Went to the U.S.A. in November 1928 with a visa allowing me to be employed there. Worked in New York City for seven months in the office of Mr. Theodore Wells, naval architect at 36 Rector Street. Following, in September 1929, received upon application a position as draftsman at the Bath Iron Works in Bath, Maine. This employment lasted for several years, I became a member of the Society of Naval Architects and Marine Engineers at that time, and during the depression in the following years had short-term jobs as designer draftsman with the Bethlehem Shipbuilding Corp. in Quincy, Mass. and the Rice Brothers Corp. in East Boothbay, Maine. The latter involved preparing construction and detail plans for a 90-foot fireboat, the City of Portland, Me. For the latter job I had moved back to Maine, the Boothbay region. I had developed a strong liking for the coast of Maine, the then rural countryside, small towns, the boatyards and most important, had found many good friends in that region. I became a U.S. citizen during the summer of 1936.

From May of 1935 to the fall of 1939 (on recommendation of G. Bear Connard, then vice president at BIW) I found employme nt with W. Starling Burgess, internationally known naval architect and engineer, as his draftsman and chiefdraftsman in his offices in Bath and Wiscasset, Maine. During those years, with much work for the Aluminum Co. of America besides yachts and commercial vessels, I took part in putting together the working drawings for the 1937 America's Cup winner "Ranger", which then was built at the Bath Iron Works.

In the fall of 1939 I built from my own designs at the yard of the Rice Brothers Corp. in East Boothbay, Maine a 26-foot auxiliary sloop, constructed of aluminum alloys, the "Whistler". This was a first, to which ALCOA gave the material. I owned that boat for 30 plus years.

Hendel's 1989 résumé written when he was 86 years old (continues on the following page).

Manuscripts Collection, Collection 193, G.W. Blunt White Library, Mystic Seaport Museum, Inc.

In 1940 I worked for a few months for Sparkman & Stephens, Inc. in New York, then moved to Ipswich, Mass. to work there in the design office of the William Robinson Shipyard for about two years, a job arranged already as far back as 1939. The most work there was on 110-foot subchasers and small landing barges, the first of wood, the last of steel construction.

In September of 1942 I was approached by friends from Wiscasset, who had connections to the Camden Shipbuilding Co. in Camden, Maine, and following went to that firm in Camden, to be in charge of the hull department there. The most work was on the 165-foot ATR rescue tugs, wood construction with steam engines, of which 12 ships were built. This job ended during the summer of 1945.

With no new job opportunities in sight and, for the past 20 years since college having worked as an employee, thus had a chance to gain experience in my field, shipbuilding in general, as well as having become known to others interested in the same field, I decided to open my own office for design services, naval architecture, in Camden, Maine.

For this, to start with, I was able to rent a room in the office building of the Camden Shipbuilding Co. The latter firm had decided to go into the business of building yachts. This development proved valuable for both parties. From the beginning I had as much work on yachts for the C.S.Co. as for outside clients. Among the latter were the Utilities and paper manufacturers, requiring small steel tugs for inland logging operations, as well as fishing vessels of various sizes and yachts, where straight power dominated. Many were built by Malcolm Brewer of the C.S.Co., which was a great help and privilege. Outstanding were the 60' sword (sport) fisherman "Porpoise", a fast powerboat built by the C.S.Co. and the 110' yacht (wood) "Vedersein" built by A & R in Germany, both ships Allison Laytham. In the beginning I hired some draftsmen, up to four at one time, young men released from the services. I best rememberRobert Jones, who was with me for 2 to 3 years, and following became a successful engineer - businessman in Massachusets. But none could be interested in boats sufficiently, would become a partner, as hoped.

Early in the 60th our office building cought fire twice, and after the second time I moved my office to 30 Bayview Street in Camden, where I continued designing until the very late 70th. Among those jobs there was the 120' yacht (steel) "Claybeth IV", built by Abeking & Rasmussen in West-Germany. From 30 Bayview Street I moved to 144 Bayview Street, a small retirement office, formerly part of our garage. A few more years there and it had been well over fifty behind the drawing-board, with the last thirty managing my own firm. I found my work rewarding, not a burden, met interesting people and also made friends doing it.

Hendel's résumé, p. 2.

Manuscripts Collection, Collection 193, G.W. Blunt White Library, Mystic Seaport Museum, Inc.

Spline Weights

Spline weights typically weigh about 3.5 lbs. and are made with lead with a felt bottom. From their physical appearance, they were called "whales" and "ducks," although "ducks" seem to have been the most common term.

As shipbuilding evolved from a craft to a science, drawings replaced full-scale wooden templates. The practice of plotting the patterns for ribs, keels, and hulls then became known as lofting, and the long curves were scribed onto the timber from a thin flexible strip of wood, steel, or plastic called a spline. The spline is bent and held in place on a flat surface by a series of several of these ducks with metal hooks that help to form a smooth and attractive curvature for draftsmen to scribe.

The ducks shown at left belonged to Geerd Hendel but were an extra set that he gave to George I. Dodge Jr., who worked as a draftsman under Hendel's supervision about 1944–47. Dodge had studied architectural drawing at the University of Maine. Edward A. Dodge Collection.

The Good Old Days of Hand Drafting, Part II, *blog by Robert Perry, November 21, 2011.*

APPENDIX III

Half-Hull and Boat Models

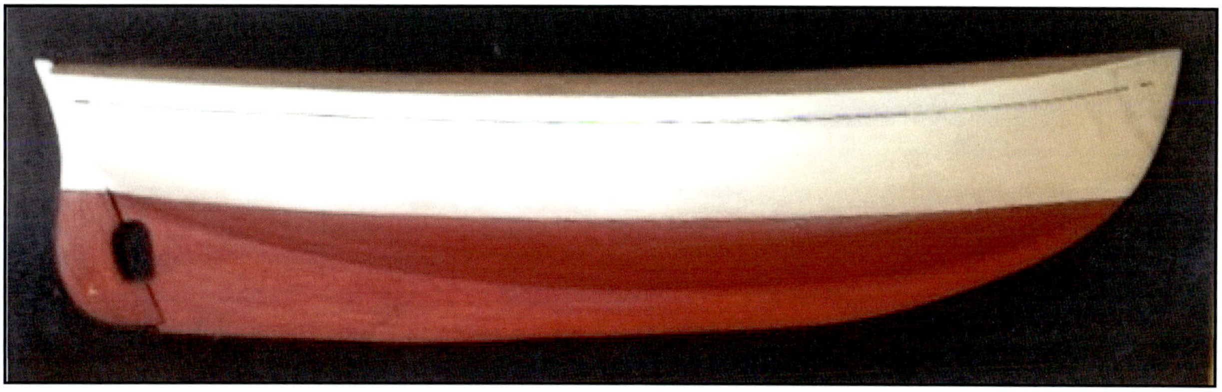

33' cruising auxiliary ketch designed by Geerd Hendel, Boothbay Harbor, Maine, 1932. Scale ½" = 1'. Model by G. Hendel

Barthold and Ingrid Brümmer Collection.

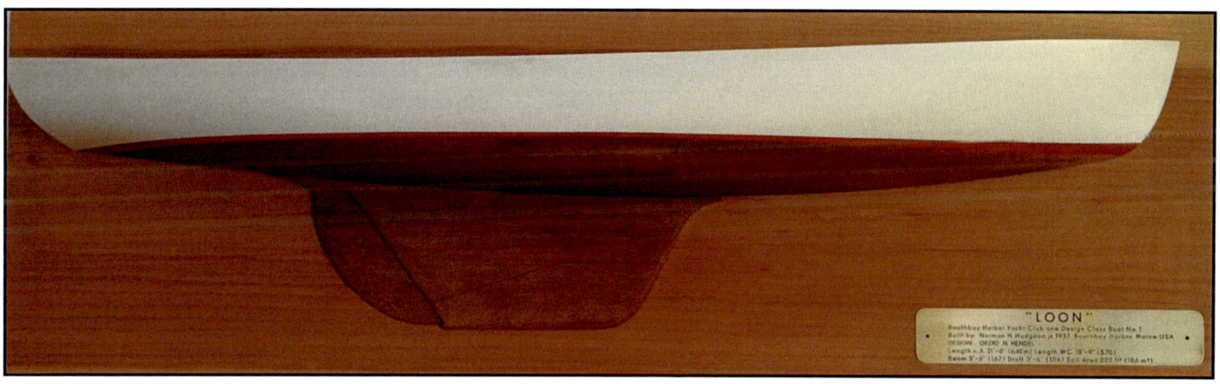

Loon.

Barthold and Ingrid Brümmer Collection.

Barthold and Ingrid Brümmer Collection.

1938: 21' Boothbay Harbor One Design (boat model builder unknown, but possibly was a Hodgdon Brothers employee who made models as a hobby).

Geerd Hendel Collection, courtesy of his stepdaughter, Brigitte M. Gefrich.

1952: 61' sport fisherman Porpoise *by boat model builder Rob Eddy.*

Geerd Hendel Collection, courtesy of his stepdaughter, Brigitte M. Gefrich.

ATR #7, one of the twelve 165' ATR salvage rescue tugs which were built for the U.S. Navy at Camden Shipbuilding during 1943 and 1944. Model by J.W. Hunt, Southport, Maine.

Barbara Dyer Collection, Walsh History Center, Camden Public Library.

Half-hull model for ATR #4, Camden Shipbuilding Co., 1943.
Ellen Stevens Norton Collection.

Ellen Stevens Norton reports that her father, James "Jim" Stevens, was working for his father, Jacob Stevens, at his shipyard, Goudy & Stevens in East Boothbay, when in 1941 Geerd Hendel suggested to Jacob that he send Jim to Camden to help establish Camden Shipbuilding's yard. Jim became the head of the lofting process for the yard, and this model was created during his two years there. It's not known whether Jim built the model himself, but he certainly handled it, refined it, and worked with it and others for two years.

ATR #4 half-hull model showing strake detail.
Ellen Stevens Norton Collection.

1952 trawler, the 92' Pocahontas*; model by Jay Hanna.*
Maine Maritime Museum.

1952 trawler Pocahontas*: model by Jay Hanna.*
Maine Maritime Museum.

Jay Hanna's model of the fishing trawler Albatross.

Smithsonian Institution Archives, Museum of American History, Maritime History section.

Half-hull model of a schooner, possibly Elmer Collemer's 36' Serenity *built in 1963, or Murray Peterson's 42' topsail schooner* Silver Heels *built by Camden Shipbuilding in 1964.*

Geerd Hendel Collection, William W. Cross (professional and news photographer, Camden-Rockland, Maine) photograph; courtesy of his stepdaughter, Brigitte M. Gefrich.

Yacht Vedersein.

Hendel Photograph Collection, Maine Maritime Museum.

APPENDIX IV

Paintings and Drawings of Hendel's Design Work

The 1938 Lightship Cornfield *was designed by the U.S. Lighthouse Service. Although not a Hendel design, Hendel worked extensively on her construction, and this painting by Robert Rice was too interesting to pass up.* Robert Rice Collection.

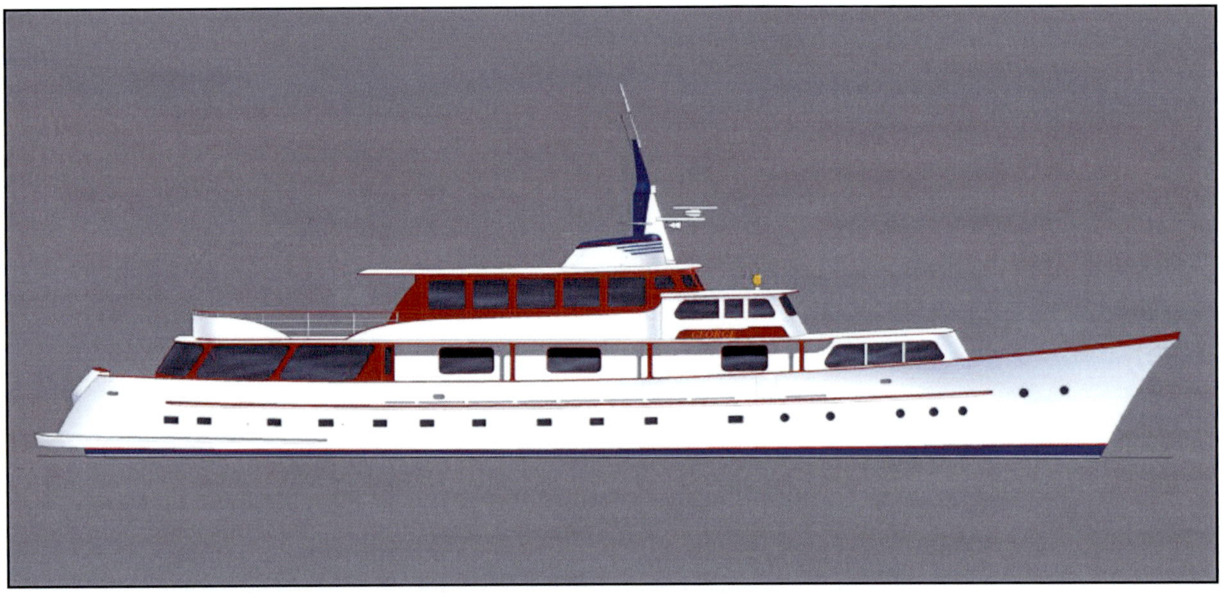

George *(formerly* Claybeth) *concept rendering showing the addition of a second deck.*

70' diesel yacht Anihita IV *(1951) by marine artist John Austin Taylor.*

Barbara Dyer Collection, Walsh History Center, Camden Public Library.

68' diesel yacht Sooner *(1954; later* Voyageur, Lady Ferrell*); artist unknown.*
Barbara Dyer Collection, Walsh History Center, Camden Public Library.

APPENDIX V

Geerd Hendel's Collection of Prints and Etchings

That Hendel kept the following image of the 1938 America's Cup race in his personal collection is certainly an outcome of his work in helping draft the designs for *Ranger*.

America's Cup race 1938 (unattributed). Endeavor II *and* Ranger.

The images that follow, primarily the work of German artist Georg Wolters, speak to Hendel's ethnic and cultural heritage, evoking his interest in and appreciation for the designs of traditional, functional, working watercraft. Those are particularly reflected in his designs for vessels such as draggers, trawlers, tugs, and boom jumpers.

Georg Wolters

Georg Wolters, whose formal name was Johann Joachim Georg Wolters, was born on May 25, 1866, in Hamburg, Germany, and died on March 11, 1943. He was a student of Gerd Hardorff the Elder and

Friedrich Schwinge in Hamburg and trained in 1890 in Paris at the Académie Rosi and in 1904 at the Academy in Antwerp. During World War I, he was a lieutenant. He was a famous yacht sailor and power boater and the only marine painter with a captain's license. His works include landscapes and seascapes with ships, shipyards, and river scenes, oil paintings, watercolors, drawings, and etchings. The printing plates were destroyed in 1943 during the bombing of Hamburg.

Geo. Wolters, 1928.

Fineknwerder Kutter Im Hafen (Fishing Cutter in Harbor *[Cuxhaven])* *in Germany on the North Sea at the mouth of the Elbe River.*

Geo. Wolters, 1925.

Gaffelschooner, Glückstadt (Gaff Schooner, Glückstadt). *Glückstadt, a town in Germany located on the Lower Elbe at its confluence with the small Rhin River, is a part of the Hamburg Metropolitan Region.*
Geo. Wolters, 1928.

Krabben Kutter at Cuxhaven. *Such vessels fished for krabben, a small North Sea shrimp.*
Geo. Wolters, 1928.

Ship in Bremerhaven Harbor (artist unknown). Bremerhaven is located at the mouth of the Weser River on the North Sea.

Gordon Hope Grant print.

Gordon Hope Grant, a noted American artist, was well known for his maritime watercolors and his work with the American Boy Scouts. He was born in San Francisco in 1875 and died in 1962.

His best-known work is likely his watercolor of the USS Constitution. *He also produced war time posters during WWI, illustrations for books such as* Penrod, *magazine covers for periodicals such as* The Saturday Evening Post, *and illustrations for* Boy's Life. *He was the cover designer for the first edition of the* Boy Scout Handbook *in 1911.*

Grant was also the illustrator for The Story of American Sailing Ships *by Charles S. Strong,* The Scarlet Plague *by Jack London,* Eternal Sea: An Anthology of Sea Poetry *edited by William Martin Williamson, and many other works.*

Wikipedia.

Source Notes

1. Wikipedia.
2. Ibid.
3. Geerd Hendel oral history interview by Joel White, Maynard Bray, and Fred Calabretta, December 14, 1988, call number OH-88-17, Item ID 006149, sound archive shelves, Marine Historical Association, Inc., G.W. Blunt White Research Library, Mystic Seaport Museum, Mystic, CT.
4. Ibid.
5. Del Sol Knight, Lucia, and MacNaughton, Daniel Brice. *The Encyclopedia of Yacht Designers*: W.W. Norton & Company, New York (2006), p. 209.
6. Ibid. pp. 209–210.
7. Howland, Llewellyn III. *No Ordinary Being: W. Starling Burgess: Inventor, Naval Architect, Aviation Pioneer, and Master of American Design*: David R. Godine, Jaffrey, NH, in association with the New Bedford Whaling Museum and Mystic Seaport Museum (2015), p. 245.
8. 1988 oral history interview with Geerd Hendel.
9. Ibid.
10. Caldwell, Bill. *Rivers of Fortune, Where Maine Tides and Money Flowed*: Down East Books, Camden, ME (1983), p. 42.
11. Snow, Ralph Linwood. *Bath Iron Works: The First Hundred Years*: Maine Maritime Museum, Bath, ME (1987) pp. 261–271.
12. 1988 oral history interview with Geerd Hendel.
13. Stephens, Olin J. II. *All This and Sailing, Too*: Mystic Seaport Museum, Inc., Mystic, CT (1999), p. 98.
14. www.atlanticclass.org/boat.
15. www.soundingsonline.com.boats. "W. Starling Burgess: A Man of Prodigious Talents and Controversies": July 31, 2015, pp. 1–2.
16. W. Starling Burgess Papers, Marine Historical Association, Inc., Manuscripts Collection 193, G.W. Blunt White Research Library, Mystic Seaport Museum.

17. Levy, Tedd. "*Cornfield* Lightship now Grand and Grounded": www.shorelandtime.com, October 12, 2011.
18. www.splinterfleet.org.
19. Wikipedia.
20. Snow, Ralph Linwood, p. 289.
21. Howland, pp. 319 and 325.
22. Snow, Ralph Linwood, pp. 290–291.
23. Ibid., p. 291.
24. Howland, pp. 320–321.
25. Ibid.
26. 1988 oral history interview with Geerd Hendel.
27. Ibid.
28. Snow, Ralph Linwood, p. 291.
29. Marine Historical Association, Inc., Manuscripts Collection 193, *Alumette* (all-aluminum hull section built at Bath Iron Works for testing durability and aluminum hulls). Calculations, photos, reports, correspondence, etc.; 1935–38, Box 8, folders 1 and 2 online, G.W. Blunt White Research Library, Mystic Seaport Museum.
30. Stephens, p. 98.
31. Howland, pp. 322–323.
32. Ibid., p. 327.
33. Ibid., p. 319.
34. Analytical report at the Maine Maritime Museum's Hendel archives, dated January 18, 1938
35. W. Starling Burgess Papers, Marine Historical Association, Inc., Manuscript Collection 193, G.W. Blunt White Research Library, Mystic Seaport Museum
36. 36 Snow, Ralph Linwood, p. 293.
37. Young, M.G., *Yachting*, November 1939, pp. 35–36
38. 1988 oral history interview with Geerd Hendel.
39. Del Sol Knight, et al., 209.
40. *Yachting*, September 1947, pp. 44–45.
41. Goudy & Stevens file, Box #1, 1920s–1990s, Boothbay Region Historical Society.
42. Howland, pp. 315–316.
43. Welles, Pete Jr. "BHYC One-Designs Play Major Role in Growth of Sailing Races Here": *Boothbay Register*, Boothbay Harbor, ME, August 22, 1952.
44. W. Starling Burgess Papers, Marine Historical Association, Inc., Manuscript Collection 193, G.W. Blunt White Research Library, Mystic Seaport Museum.
45. Howland, p. 315.

46. *Waterline,* Newsletter of the Boothbay Harbor One Design Association, December 2007, p. 6.
47. Foster, Robert C. III, and Reed, Alden II. "The Boothbay Harbor One Design" in Barbara Meeks Chapman's *A History of the Boothbay Harbor Yacht Club from 1895 to 2007,* June 2007, Boothbay Harbor, ME, self-published.
48. Welles article in the Boothbay Register, August 22, 1952.
49. Ibid.
50. Howland, p. 316.
51. Ibid.
52. *Waterline,* Newsletter of the Boothbay Harbor One Design Association, Vol. 1, Issue 2, Summer 2005, pp. 1 and 5.
53. Chapman, Barbara Meeks. *A History of the Boothbay Harbor Yacht Club from 1895 to 2007,* June 2007, Boothbay Harbor, ME, self-published.
54. BHODA 2010-2011 Membership Directory p.20
55. *The Boothbay Harbor One Design by Geerd N. Hendel,* The Artisan Boatworks, Rockport, ME (www.artisanboatworks.com).
56. Duncan, Roger S. *The Working Waterfront* magazine, July 1, 2007.
57. Swan, Debbie. "David Nutt Discusses Building New One Designs": report in *The Wiscasset* Newspaper re: meeting of the Boothbay One Design Association at Juniper Point, August 24, 2013, posted on Friday, August 23, 2013.
58. Bray, Maynard, in *WoodenBoat* magazine's 2007 *Small Boats* Annual, p. 72.
59. Chapman, Barbara Meeks.
60. Rumsey, Barbara, Boothbay Region Historical Society curator. "Jim Stevens Notes on One-Designs," an interview dated May 16, 2002.
61. "History of the Boothbay Harbor One Design," online at www.learn-to-sail.org/bhodhistory.
62. Howland, p. 316.
63. *Waterline,* Newsletter of the Boothbay Harbor One Design Association, Volume 2, Issue 2, Summer 2006, p. 2.
64. Ibid., pp. 7–8.
65. Howland, p. 317.
66. Ibid., p. 337.
67. 1988 oral history interview with Geerd Hendel.
68. www.yankeeonedesign.com/about-the-yankee-one-design-class.
69. Howland, p. 341.
70. Stephens, p. 148.
71. Clifford, Harold B. *The Boothbay Region, 1906–1960,* The Cumberland Press, Freeport, ME (1982), p. 261.

72. Del Sol Knight, et al., p. 209.
73. Hendel, Geerd, letter to S.B. Foot, November 9, 1935: Geerd Hendel Collection, Maine Maritime Museum, MS-310, 38.106.1, file #2.
74. Stephens, pp. 97–99.
75. Ibid., p. 148.
76. *Boating* magazine, July–December 1966, p. 38.
77. Wyland in *Yachting* magazine, September 1947.
78. Wikipedia.
79. www.esbna.com/us/en/education/blog/a-short-history-of-welding-aluminum.cfm.
80. *Yachting*, September 1939, pp. 35–36
81. *MotorBoating* magazine, September 1953, p. 146.
82. ttps://www.yachtcharterfleet.com/luxury-charter-yacht-25452/george.html.
83. Dyer, Barbara. *Grog Ho! The History of Wooden Vessel Building in Camden, Maine: Courier-Gazette*, Inc., Rockland, ME (1984), pp. 143 and 147.
84. Ibid.
85. Wikipedia.
86. Snow, Bertram B. *The Main Beam*: Rockland Historical Society, Rockland, ME (2005), p. 169.
87. Snow, Ralph Linwood, p. 301.
88. Dyer, pp. 90–93.
89. Ibid., p. 143.
90. Ibid.
91. Ibid., p. 153.
92. Ibid., p. 156.
93. Ibid., p. 161.
94. Ibid., p. 154.
95. Ibid., p. 161.
96. "Design by Geerd Hendel": commentary by Joel White-Boat Designs, Thursday, October 24, 2013, http://www.boatbuilding.xyz/boat designs.
97. Dyer, p. 164.
98. Ibid., p. 166.
99. *The Portland Press Herald*, April 9, 1957, Walsh History Center, Camden Public Library.
100. Dyer, p. 168.
101. Ibid., p. 172.
102. Snow, Bertram B., p. 458.
103. Ibid.
104. Ibid.
105. Ibid., p. 460.

106. Ibid., p. 462.
107. Gould, John, *Christian Science Monitor*, 1993.
108. *Bangor Daily News*, September 5, 1972.
109. *The Portland Press Herald*, February 20, 2016.
110. *The Maine Citizen*, January 16, 1992.
111. *The Northern*, magazine of the Great Northern Paper Company, September 1928, p. 3: Great Northern Paper Company Records, SpC MS 0210, Box 21, Raymond H. Fogler Library Special Collections Department, University of Maine, Orono, ME.
112. Snow, Bertram B., p. 456.
113. Ibid., p. 378–380.
114. MacNaughton, Dan. "The Light Displacement Controversy" in 61/*WoodenBoat*, pp. 42–43.
115. Editors of *WoodenBoat* magazine, *50 Wooden Boats: A Catalog of Building Plans*: WoodenBoat Publications, Brooklin, ME (2003).
116. Day, Jane, 49/*WoodenBoat*, p. 49.
117. Dyer, pp. 128–129.
118. Dyer, p. 129.
119. Day, 49/*WoodenBoat*, p. 49.
120. Geerd Hendel Collection, MS-310, Box 3, Maine Maritime Museum.

Index